NUCLEAR ENERGY AND INFORMATION

RADIATION IN PERSPECTIVE

Applications, Risks and Protection

NUCLEAR ENERGY AGENCY
ORGANISATION FOR ECONOMIC CO-OPERATION AND DEVELOPMENT

ORGANISATION FOR ECONOMIC CO-OPERATION AND DEVELOPMENT

Pursuant to Article 1 of the Convention signed in Paris on 14th December 1960, and which came into force on 30th September 1961, the Organisation for Economic Co-operation and Development (OECD) shall promote policies designed:

— to achieve the highest sustainable economic growth and employment and a rising standard of living in Member countries, while maintaining financial stability, and thus to contribute to the development of the world economy;

— to contribute to sound economic expansion in Member as well as non-member countries in the process of economic development; and

— to contribute to the expansion of world trade on a multilateral, non-discriminatory basis in accordance with international obligations.

The original Member countries of the OECD are Austria, Belgium,Canada, Denmark, France, Germany, Greece, Iceland, Ireland, Italy, Luxembourg, the Netherlands, Norway, Portugal, Spain, Sweden, Switzerland, Turkey, the United Kingdom and the United States. The following countries became Members subsequently through accession at the dates indicated hereafter: Japan (28th April 1964), Finland (28th January 1969), Australia (7th June 1971), New Zealand (29th May 1973), Mexico (18th May 1994), the Czech Republic (21st December 1995), Hungary (7th May 1996), Poland (22nd November 1996) and the Republic of Korea (12th December 1996). The Commission of the European Communities takes part in the work of the OECD (Article 13 of the OECD Convention).

NUCLEAR ENERGY AGENCY

The OECD Nuclear Energy Agency (NEA) was established on 1st February 1958 under the name of the OEEC European Nuclear Energy Agency. It received its present designation on 20th April 1972, when Japan became its first non-European full Member. NEA membership today consists of all OECD Member countries, except New Zealand and Poland. The Commission of the European Communities takes part in the work of the Agency.

The primary objective of NEA is to promote co-operation among the governments of its participating countries in furthering the development of nuclear power as a safe, environmentally acceptable and economic energy source.

This is achieved by:

— *encouraging harmonization of national regulatory policies and practices, with particular reference to the safety of nuclear installations, protection of man against ionising radiation and preservation of the environment, radioactive waste management, and nuclear third party liability and insurance;*

— *assessing the contribution of nuclear power to the overall energy supply by keeping under review the technical and economic aspects of nuclear power growth and forecasting demand and supply for the different phases of the nuclear fuel cycle;*

— *developing exchanges of scientific and technical information particularly through participation in common services;*

— *setting up international research and development programmes and joint undertakings.*

In these and related tasks, NEA works in close collaboration with the International Atomic Energy Agency in Vienna, with which it has concluded a Co-operation Agreement, as well as with other international organisations in the nuclear field.

Publié en français sous le titre :
LE POINT SUR LES RAYONNEMENTS
Applications, risques et protection

© OECD 1997

Cover: Model of the repository for low- and medium-levels radioactive waste at Olkiluoto, Finland.
Credit: TVO, Finland.

DEDICATION

This report is dedicated to the memory of Henri Jammet, one of the pioneers of radiation protection and founder of radio-pathology, who died unexpectedly on 19 August 1996, aged 76. He had originally planned to write this book but his declining health prevented him from carrying out this task.

Henri Jammet started as a physician at the French "Institut du Radium" (now "Institut Curie"), where he became the head of the Department of Radio-pathology, which was created in 1956 at the request of Irène Joliot-Curie to care for radiation victims. He became known to the general public when he co-ordinated the treatment of several irradiated casualties, especially those resulting from the accidents in Vinca (1958), Mol (1965), Setif (1978) and Casablanca (1984). He rapidly developed a dual career and, as Director for Protection at the Institute for Protection and Nuclear Safety, he was able to initiate intensive research programmes on the health effects of radiation.

As his experience covered an extremely wide area, encompassing all aspects of the protection of man against ionising or non-ionising radiation, he was elected in 1953 to membership of the International Commission on Radiological Protection (ICRP) of which he became Vice-Chairman in 1985. In addition, he belonged to several national and international scientific organisations, including the United Nations Scientific Committee on the Effects of Atomic Radiation, where he chaired the French delegation.

One of his essential merits was his early recognition of the importance of international dialogue as a main source of renewal and thorough knowledge. Specialists of several disciplines trusted him and followed his guidance. His ability and personal commitment, the precision of his reasoning, his desire to serve, his legendary devotion to his patients and his unflinching optimism were recognised world-wide.

He was always warm and friendly; the curiosity of his spirit was never satisfied, and this is why, throughout his life, he sought to improve his overall understanding of human nature and the environment. All who had the fortune to know him will always remember his engaging personality, his human warmth, his simplicity, his lack of pretention, his humanity and his culture. He leaves a large vacuum which will be very hard to fill.

Acknowledgements

The NEA Secretariat wishes to express its sincere appreciation to Mr. Peter Saunders and Dr. Jean-Claude Nénot for their invaluable contribution to the preparation of this report. Mr. Saunders is a technical writer and a consultant on energy and environmental issues. He is a Partner in Technology Business Communications. Dr. Nénot is a specialist of radiobiological protection of workers and population and is Director for Research at the Institute for Protection and Nuclear Safety of the French Atomic Energy Commission.

The NEA Secretariat also wishes to thank the Japanese authorities for their financial support towards the publication of this report.

FOREWORD

This report is concerned with ionising radiation. Despite its many beneficial applications throughout medicine, industry and research, ionising radiation has led to much public controversy. The issue sprang to prominence in the immediate aftermath of the Hiroshima and Nagasaki bombings, and it has played a large part in the debate about the generation of electricity by nuclear power.

Although unknown just a hundred years ago, ionising radiation is hardly a new phenomenon. Mankind has always been exposed to radiation from outer space and from radioactive materials in the earth itself. But it is man-made radiation that gives rise to the greatest concern.

A fundamental requirement, both in developed and in developing countries, is to enable the benefits of technology to be enjoyed without unacceptable consequences. The debate about radiation is just one of many similar debates that have arisen in countries, such as the Member countries of the OECD, in which high standards of living depend on a wide range of technologies, applied through a complex infrastructure of manufacturing and energy industries and transport systems. The challenge is to maintain or improve these standards without undue damage to the environment, and without compromising the ability of future generations to meet their own needs – to achieve sustainable development, as defined in the 1987 report of the World Commission on Environment and Development "Our Common Future", the Brundtland Report.

The controversy about radiation is based on a series of apparent paradoxes: it is universal, yet unseen and unfelt; it has many beneficial applications, yet can be harmful; it is well understood, yet often feared. Measures to safeguard health and the environment from the harmful effects of radiation are well developed and applied through stringent national legislation in most countries. There is a broad scientific and technical consensus that the degree of scientific knowledge of radiation and its effects constitutes an acceptable basis for a conservative system of protection. Yet the debate continues.

This report aims to contribute to this debate by summarising the most up-to-date and authoritative material that is available on the sources and effects of radiation and on the ways in which people are protected from its risks.

The report looks in detail at the way radiation and radioactive materials are applied in a wide range of medical, industrial and research activities as well as in energy production; the radiation exposures that result from these activities and the generally much larger radiation exposures from natural sources; and the effects of radiation on living matter, particularly on human health.

It discusses the development of radiation protection measures, the internationally agreed principles on which protection measures are now based throughout the world, and the roles and activities of the national and international bodies involved in making recommendations and setting and enforcing standards for safety.

Finally, it addresses social and economic issues such as ethical questions, risk perceptions, risk comparisons, public understanding and participation in decision-making, and the costs of protection, and sets out some possible future problems, opportunities and topics for further research.

It is produced by the OECD Nuclear Energy Agency, whose primary objective is to promote co-operation among the governments of its participating countries in furthering the development of nuclear power as a safe, environmentally acceptable and economic energy source. Public understanding and acceptance of all the activities associated with such development is a key requirement in all of these countries. The Agency is seeking to achieve this through the provision of a series of comprehensive reports on key issues in a form that is useful to government circles and readily understandable by the interested non-specialist public. This report is one of that series.

The opinions expressed in this report do not necessarily represent the views of any Member country or international organisation. It is published on the responsibility of the Secretary-General of the OECD.

TABLE OF CONTENTS

Chapter 1

INTRODUCTION

Early discoveries

The foundations of today's extensive understanding of ionising radiation (referred to in the rest of this report as radiation – see box on page 12) were laid during a short period of brilliant research, starting with Röntgen's discovery of X-rays in November 1895. Röntgen was working in a darkened laboratory, studying what happened when an electric current passed through a glass tube out of which most of the air had been pumped. The current made the remaining air in the tube glow. He covered the tube with black cardboard, so that the glow could not be seen. A sheet of paper that had been made sensitive to light happened to be lying on the table and Röntgen found that this paper glowed every time the current flowed, although no light was escaping through the cardboard. He then found that whatever was coming from the tube also darkened a photographic plate. He realised that some sort of radiation invisible to the human eye was being emitted and he called this X-rays. He soon found that X-rays could be stopped by some substances more easily than by others. One of his most dramatic early demonstrations was to put his hand between the tube and a fluorescent screen; the image on the screen clearly showed up the bones. Within a very short time, X-rays were being put to medical use, first to locate shotgun pellets in a man's hand and soon for a very wide range of diagnoses in hospitals and surgeries throughout the world.

In 1896 Becquerel found that pitchblende, an ore of uranium, caused the darkening of a photographic plate; he had discovered radioactivity. Rutherford identified two types of radiation being emitted from pitchblende, neither of which were visible to the human eye; he called these alpha and beta rays. He showed that the alpha rays (or alpha particles) consisted of nuclei of the element helium. Becquerel showed that beta rays (or beta particles) were electrons. Villard found a third type of radiation emitted from pitchblende which he called gamma rays. These were later shown to be electromagnetic radiation similar to, but more penetrating than X-rays. In 1896 Pierre and Marie Curie extracted a minute amount of radium – less than one gramme – from many tonnes of pitchblende, and radium was soon being used in cancer treatment because of the intense radiation it emitted.

Thus by the beginning of the twentieth century many of the key discoveries needed to understand man's radiation environment had been made, and the stage was set for the development of applications of radiation in medicine, industry and research.

IONISING RADIATION

Strictly speaking, the term ionising radiation should be used when describing X-rays and alpha, beta and gamma rays, to differentiate them from more familiar types of radiation, such as light, heat, radio waves, and microwaves. Ionising radiation is so-called because it produces an electrical effect called ionisation when it interacts with matter. As the radiation passes through matter it disrupts some of the atoms in its path, leaving a trail of electrons and charged atoms or ions. The process is important because these free ions can cause physical and chemical changes to take place in the surrounding material. These changes can affect its properties; in particular, they can cause harm in living matter.

Modern alchemists

A second dramatic series of discoveries was initiated in 1919 when Rutherford fired alpha particles from a radium source down a tube containing nitrogen gas. At the end of the tube he detected protons as well as alpha particles. Protons are the nuclei of hydrogen, but there was no hydrogen in the tube. What was happening in Rutherford's experiment was that some of the alpha particles were being captured by the nitrogen nuclei, and the subsequent nuclear rearrangements resulted in the formation of oxygen and the emission of protons. One element had been transmuted into another, the alchemist's dream, only in this case nitrogen was changed into oxygen rather than lead into gold.

In 1932, Cockcroft and Walton showed that nuclear transmutations could be caused by protons accelerated by electrical means in the laboratory as well as by alpha particles from naturally radioactive materials. In 1934, Irene Curie and Frederic Joliot found that in some circumstances transmutation resulted in the formation of new elements that were radioactive (a radioactive element is called a radionuclide). They had discovered artificial, or man-made radioactivity. The next year, Fermi found that a much greater variety of artificial radionuclides could be created when neutrons were used instead of alpha particles or protons. Neutrons – particles of approximately the same mass as protons but electrically neutral, present in the nuclei of all elements except hydrogen – had been identified by Chadwick in 1932.

Although the quantities of artificial radionuclides produced in these various ways were minute, the radiation they emitted could easily be detected, and this enabled them to be used as tracers (small amounts of easily detectable material fixed to some substance whose movement one wishes to follow). The use of radioactive tracers, a technique invented by Hevesy, has since revolutionised our understanding of the working of the human body, in health and in sickness. A radionuclide behaves chemically in exactly the same way as the non-radioactive form of the same element. So adding a minute amount of a radioactive tracer to a chemical and measuring the radiation it emits allows the movement and metabolism of the chemical to be studied with enormous sensitivity, typically over a million times greater than that attainable by other means. Radioactive tracers are also widely used in industry and research as well as in medicine (most are now produced in nuclear reactors).

From 1934 onwards, Fermi and other researchers irradiated a wide range of elements with neutrons, including the heaviest known element, uranium. For most elements, the absorption of a neutron results in the formation of a heavier element. In the case of uranium, however, a wide range of elements results, some of which are heavier but others much lighter than the original uranium. The absorption of the neutron can in fact cause the nucleus to split into two (or sometimes more) fragments, most of which are radioactive. The process is called fission. Fission had been occurring in the experiments of Fermi and others but this was not realised until Hahn, Strassmann and Meitner in 1939 showed that one of the new elements formed was barium, an element about half the weight of uranium.

Nuclear power

The characteristics of nuclear fission were soon established as a result of a series of further experiments and theoretical studies. A key feature is that the fission products contain fewer neutrons than the uranium from which they originate, the surplus neutrons being emitted during the fission process. There was clearly the possibility of achieving a chain reaction in which some of the

13

emitted neutrons induce fissions in other uranium nuclei, resulting in more neutrons, more fissions, and so on. Furthermore, the mass of the fission products plus the neutrons was less than the mass of the original uranium, with the difference in mass (m) appearing in the form of the kinetic energy (E) of the fission fragments and neutrons, in accordance with the Einstein relationship $E=mc^2$ (c being the velocity of light), leading to the possibility of releasing enormous amounts of energy. This is the only formula that will be found in this report.

The possibility of a self-sustaining nuclear chain reaction was first confirmed by Fermi on 2 December 1942, using an assembly (or pile) consisting of uranium, graphite which slowed the neutrons emitted from the fission reactions down to the optimum speed for initiating further fission reactions, and cadmium which controlled the chain reaction by absorbing neutrons. This, the first nuclear reactor, was constructed in a squash court under the stand of a baseball stadium in Chicago. The possibility of an enormous energy release was confirmed at the Alamogordo Bombing Range in New Mexico on 16 July 1945 when the first nuclear weapon was tested. Both these events were shrouded in the strictest secrecy. The world first learned about the power of the nucleus at Hiroshima and Nagasaki.

The Chicago reactor showed that the power of nuclear fission could be produced under controlled conditions. Many reactor types were developed during the subsequent decades. All are based on assemblies of fissile materials, uranium and plutonium in various forms, together with means of controlling the numbers of neutrons and, if necessary, their speed, and of removing the energy released and using it to generate electricity. Plutonium is one of the new elements that result from the absorption of a neutron by uranium; it is another element in which a neutron chain reaction can be sustained and was the material used in the Nagasaki bomb.

Nuclear weapons and nuclear electricity generation have been important and contentious topics for half a century. For better or worse, nuclear weapons have dominated much military thinking and national and international defence strategies throughout the period. Nuclear electricity generation has grown steadily since the early 1960s and now provides one quarter of the electricity generated in the OECD countries, the second largest source after coal. The importance of these two radically different activities in the context of this report is in their inescapable association with radiation. Although, as will be seen in later chapters, they contribute only a small fraction of mankind's total radiation exposure, they are strongly connected in most people's minds with the "problem"

Nuclear electricity as a percentage of
total electricity supplied (1995) in the OECD area

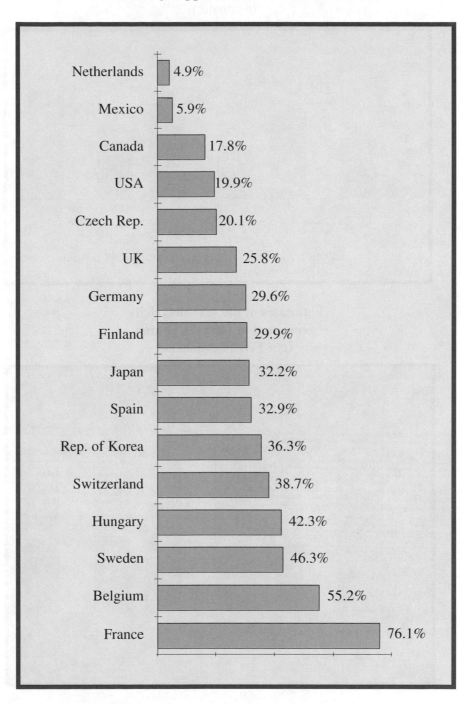

Country	Percentage
Netherlands	4.9%
Mexico	5.9%
Canada	17.8%
USA	19.9%
Czech Rep.	20.1%
UK	25.8%
Germany	29.6%
Finland	29.9%
Japan	32.2%
Spain	32.9%
Rep. of Korea	36.3%
Switzerland	38.7%
Hungary	42.3%
Sweden	46.3%
Belgium	55.2%
France	76.1%

Nuclear electricity as a percentage of
total electricity generated in the OECD area
(estimates)

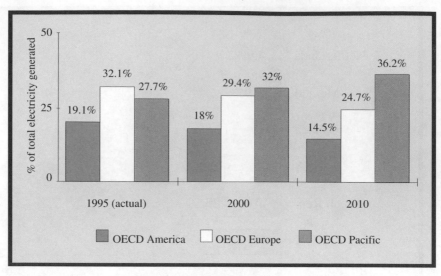

Estimates of nuclear electricity
generated in the OECD area
(net TWh) (estimates)

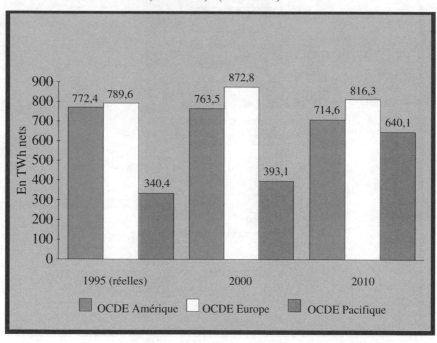

of radiation and its safe control. Attention focused initially on radiation exposures from the fallout from atmospheric testing of nuclear weapons, and later, following international agreements banning atmospheric testing, on radiation associated with nuclear electricity generation.

Risks and benefits

The possible dangers of radiation were recognised within a very short time of its initial discovery. Speaking of X-rays in 1896, Lord Lister said: "If the skin is long exposed to their action it becomes very much irritated, affected with a sort of aggravated sun-burning. This suggests that transmission through the human body may not be altogether a matter of indifference to the internal organs." Rutherford visited the Curies in 1903 and noted: "We could not help but observe that the hands of Professor Curie were in a very inflamed and painful state due to exposure to radium rays."

Several hundred cases of radiation injury were reported during the early years of the medical use of radiation, mainly among researchers, doctors and technicians – there is a monument in Hamburg to 169 radiologists who died from such injuries. It was this early experience that led to the development of measures for the protection of people from the harmful effects of radiation.

Within a few years, it was realised that, as well as causing gross damage such as skin burns, large or repeated exposures to radiation could sometimes cause cancer. Later, experiments on animals and other living matter showed that radiation exposures could produce effects which only show up in the offspring of the exposed organisms – genetic or hereditary effects.

There is a threshold level of radiation exposure (or dose) below which no acute damage is observed. However, there is now a substantial body of evidence which shows that cancers can be caused by radiation even at dose levels that are below this threshold. There is no evidence of inherited damage in humans at any dose level, but the fact that such damage is found to occur in other life forms suggests that it could also occur in humans. While the question of whether or not very low doses of radiation, well below the threshold for acute effects, result in cancers or genetic effects remains an open question, nevertheless the assumption that a dose, however small, carries with it an element of risk has become a key concept in radiation protection.

A key problem of radiation protection is that of balancing risks and benefits. There is little doubt that the benefits to patients from the use of radiation

17

and radioactive materials for diagnosis and treatment generally far exceeds any associated risks. The balance becomes more difficult if the benefits are general but the risk is concentrated on a few people. The benefits of nuclear power, for example, are general – it helps to reduce the problems of acid rain and greenhouse gas emissions from fossil fuelled electricity generation and contributes to the diversity of energy supply. But to set against such benefits there are risks to those working in the industry and to the public around nuclear installations. Such complex problems are not unique to the nuclear industry, they also arise when considering the siting of motorways, airports, chemical plants and waste disposal sites. In all these cases, the benefits are general but the risks and disadvantages are mainly local.

Further questions arise when considering issues that are essentially socio-economic rather than technical. To what extent should decision-makers take public perceptions of risk into account when there is a discrepancy between such perceptions and quantitative risk estimates derived from technical analysis of engineering systems? How much should be spent on the further reduction of risks that are already low? How should benefits to current generations be balanced against possible risks to future generations?

There are no simple answers to such questions, but they can only be addressed on the basis of a clear understanding of the technical issues, which this report seeks to provide.

Chapter 2

APPLICATIONS OF RADIATION

The benefits of radiation became apparent within a very short time of the early discoveries, particularly in medicine. Techniques based on radiation and radioactive material are now available for application in a wide variety of fields. In addition to making possible ever more precise and complex medical diagnosis and therapy, they are increasingly used in industry, food production, environmental measurement and control, archaeology, research and many other areas.

Some applications make use of radionuclides occurring naturally in the environment. Others involve the use of X-rays or the radiation from nuclear reactors or particle accelerators, either directly or as a means of producing useful radionuclides. Many of the techniques are well-known and long-established, although even these are continuously being refined and improved; others have only recently become possible as a result of new discoveries and technical developments. A few examples from the rapidly growing number of applications are given in the following sections.

Medicine

Medical applications of radiation generally fall into three categories:

- direct use in diagnosis;
- techniques based on the ability of radiation to destroy bacteria, malignant cells or other unwanted tissue;
- the use of radionuclides as tracers to investigate the functioning of specific organs.

X-ray diagnosis remains by far the most common and widespread application of radiation. Even after a hundred years of use, improvements are still being made, particularly with the advent of powerful computers that enable detailed three-dimensional pictures to be generated from a series of scans through the body. Radiations other than X-rays are also being used in similar ways.

19

Radiation can be sufficiently intense to destroy bacteria and hence it is extensively used for sterilising medical equipment. The most common source is cobalt-60; electron beams from accelerators can be used for the same purpose. At present, in the developed countries, about 45 per cent of medical products are sterilised using radiation and this is expected to grow to about 80 per cent over the next few years. Disposable medical equipment and dressings can be treated in this way to the benefit of doctors and patients, particularly in developing countries where other methods of sterilisation would be impracticable on the spot. Sterilisation by radiation is projected to grow rapidly and many plants are being built in Africa, the Middle East and South America.

Radiation therapy is widely used in many countries, most commonly in the treatment of cancer. Typically about two people per 1 000 population in developed countries receive such treatment annually, corresponding to about 100 000 patients in a country with 50 million inhabitants. During treatment, radiation from an external source is carefully focused on the tumour itself to minimise damage to surrounding tissues. For some treatments, radioactive sources are inserted into the patient, or radioactive materials are given to the patient which are subsequently carried to the cancerous organ by the bloodstream.

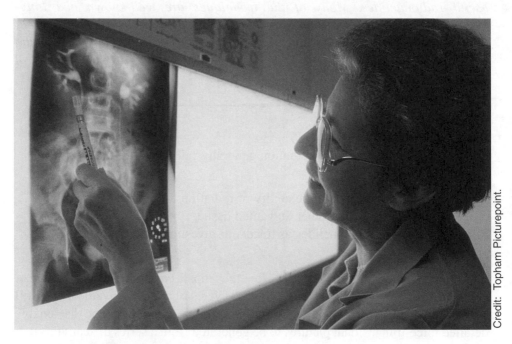

Nurse looking at an X-ray radiograph.

A wide range of radionuclides is now available for medical use as tracers. One of the most important is technetium-99m, which comes from another radionuclide, molybdenum-99, produced in nuclear reactors. Technetium-99m emits radiation which is ideally suited for detection in modern gamma ray cameras and has a short half-life of only 6 hours (half-life is defined in the next chapter), resulting in very small doses to patients. Because of its chemical versatility, it can be combined in many different compounds enabling a range of organs to be studied. Every day, world-wide, some 60 000 medical procedures are carried out that involve the use of technetium-99m.

Iodine-131, which is a common fission product from nuclear reactors, has been used for many years in medicine. The thyroid gland takes up iodine, hence any iodine-131 introduced into the body will make its way there where its presence and movement can be monitored by detecting the gamma radiation which it emits. Recently, another isotope of iodine has become available for this purpose – iodine-123. This is produced by a charged particle accelerator rather than a reactor. It can now be produced in an isotopically pure form and has many advantages over iodine-131. It has a much shorter half-life, 13 hours compared to 8 days, and unlike iodine-131 does not produce radiation other than gamma rays. As a result the radiation dose to the patient can be significantly reduced.

Another useful, though very different, radiation technique that has found medical application is ion implantation. Here a beam of heavy ions from a low energy accelerator is fired onto the surface of a host material. The energy of the ions is too low to cause nuclear reactions but sufficient for them to become embedded in the surface layer and transform its properties. Nitrogen ions implanted into the surface of the titanium alloy have been used to make replacement hip joints which are a thousand times more resistant to wear than those made with untreated alloys. This means they will last much longer with obvious advantages to the patient.

Food production

Sterilisation by gamma irradiation can be used to destroy the bacteria and insects responsible for the degeneration of food, for example replacing the use of DDT at high concentrations for the treatment of dried fish. The gamma rays kill insect larvae which can otherwise reduce the sun-dried fish to bare bones within a few weeks. Since 1993 an irradiation plant for food and medical supplies has been operating, under the International Atomic Energy Agency (IAEA) auspices, sited at the heart of the fishing region in Bangladesh and with enough capacity to treat all the dried fish the country produces.

Radiation techniques are also used to control pests. Thirty-six countries in Africa are infested by tsetse flies – a total area of 10 million square kilometres. They cause devastating effects on livestock and their eradication would have enormous benefits for social and economic development; it would also prevent the environmental damage now being done by chemical controls. The sterile insect technique has the potential to eradicate this pest and is being applied in a number of areas under IAEA auspices. Large numbers of male insects are sterilised by radiation and released, resulting in mostly infertile matings and a rapid reduction in the insect population.

There has been increased use in recent years of controlled release pesticides, where the active ingredient is enclosed in a polymeric matrix which releases it at a specified rate over a fixed period of time. This rate must be well understood and is usually measured by using radionuclides as tracers. Those most suitable for this purpose are tritium, carbon-14, phosphorus-32 and chlorine-35.

A better understanding of the uptake of fertiliser by food crops can lead to a dramatic increase in yield. One aspect of this is to find out at which time in the growing cycle of a specific crop fertiliser should be applied to maximise its effectiveness. The application of radionuclides as tracers has proved to be a useful technique in acquiring this knowledge. The most suitable radionuclides for this purpose are nitrogen-15 and phosphorus-32.

Credit: JAERI, Japan.

Irradiation is used to control sprouting of potatoes during storage.

The availability of fresh water is essential for human well-being, both for direct consumption and to irrigate crops. There is growing concern over the quantity and the quality of future supplies. For example, it is projected that early in the next century supplies in Egypt will be down to two thirds of their present level and in Kenya down to a half. Serious problems are also predicted in several other East African countries and in those on the south Mediterranean coast. In China the water table below Beijing is sinking at a rate of about 2 metres per year and about a third of the country's wells are running dry. In parts of India the water table has fallen by 25 metres in the past 20 years. Also there is degradation of supplies world-wide by pesticides and fertilisers in run-off from the land and by heavy metals and hydrocarbons in the storm water from cities.

Clearly there is a need to assess, manage and protect our water resources and radionuclides play a significant role in this process. For example tritium is very useful as a tracer for water movement and naturally occurring carbon-14 and chlorine-36 are used to determine the age of ancient aquifers from which their replenishment time can be estimated. Since 1987 the IAEA has been sponsoring the Sahel project which uses these techniques in sub Saharan Africa.

Industry

The use of radiation techniques in industry has grown dramatically over the past few decades. In Australia, for example, there were 125 devices in use in 1961; by 1990 there were over 2 000. Similar growth has taken place world-wide, taking advantage of many different properties of radiation.

Beta rays are attenuated, but not completely stopped, by paper. Hence by placing a beta source such as krypton-85, a fission product, on one side of a piece of paper and a detector on the other, a measurement of the count rate enables the thickness of the paper to be determined. This finds application in paper mills where a feedback signal from the detector can be used to control the paper thickness within prescribed limits.

Californium-252 is an actinide with a very unusual property. While it may decay by alpha, beta or gamma emission, it has an alternative mode of decay by spontaneous fission with the emission of a neutron. The yield from this mode of decay is high enough to make it a convenient, and portable, source of neutrons which finds many applications in industry, in particular for on-line prompt gamma analysis. In this technique, neutrons excite stable nuclei which then promptly de-excite emitting a characteristic gamma ray whose detection then identifies the element. This is used for on-line multi-element analysis of coal as

Radioactive materials are used to measure cylinder wear in car engines.

it is delivered to a power station. Measurement of elements such as silicon, iron, aluminium and calcium enable the ash content of the coal to be estimated, and a determination of sulphur gives warning of the extent of possible air pollution when the coal is burned. This technique is also used in the gold mining industry to determine the residual gold in the tailings, and in towed seabed spectrometry to perform a multi-element analysis of the seabed to identify possible economic mineral deposits.

The radionuclide zinc-65, a gamma emitter with a half-life of 244 days, is used in the automobile manufacturing industry to measure the oil consumption of new engine designs under laboratory conditions. A small amount of the radionuclide is added to the oil and the exhaust gas is monitored for the characteristic gamma ray. This enables an instantaneous measurement of oil consumption to be made under varying conditions of load and speed.

The intense flux of neutrons inside a reactor can be used to produce semiconductor material for the electronics industry. The requirement is to introduce phosphorus at the level of a few parts per million into silicon, and to do so on a bulk scale and with a high degree of uniformity. Natural silicon contains the stable isotope silicon 30 with an abundance of about 3 per cent.

When an ingot of silicon is placed in the core of a reactor, this isotope absorbs neutrons to produce the radionuclide silicon 31 which decays by beta emission with a half-life of 2.6 hours to the stable isotope phosphorus-31. As the neutrons permeate evenly throughout the whole of the ingot, the uniformity of phosphorus distribution greatly exceeds that achievable by other techniques. The process is ideally suited to bulk production and currently yields about 50 tonnes per year world-wide.

By providing an exit channel through the shielding of a reactor, a beam of neutrons can be produced. Such beams have many uses, one of which is neutron radiography. This is complementary to the more familiar X-radiography in that, unlike X-rays, neutrons are strongly absorbed by light elements such as hydrogen whereas a heavy element such as lead is virtually transparent. Thus neutron radiography can be used to obtain images of lighter, particularly hydrogenous, material inside a heavy metallic casing. One application is the investigation of the flow of lubricating oil inside a working engine.

Accelerators can produce X-rays of much higher energy, and hence of much greater penetrating power, than is available from conventional machines and these can be used for examining thick section components. For example this technique enables radiographic images to be obtained of the turbine blade tip clearances in jet engines whilst they are operating, through casings of up to 20 cm thickness. It can also be used to examine the condition of the steel in bridges through 80 cm thick concrete beams.

Charged particle beams from accelerators have been put to a wide range of practical uses. One industrial application is thin layer activation, which can be used to measure wear in engine parts. The surface of the critical component is bombarded by protons from an accelerator with an energy of about 12 million volts. At this energy protons penetrate into steel to a depth of about a third of a millimetre and in so doing convert some of the iron nuclei into cobalt-56, a gamma emitter with a half-life of 77 days. When the engine is running any wear on this surface will result in some of the cobalt-56 passing into the lubricating oil. Subsequent measurement of its characteristic gamma radiation from the oil enables the rate of wear to be determined.

Electron beams from accelerators can be used to produce cross-linked polymers which have many useful properties such as much improved electrical and thermal insulation. Such materials are put to a wide range of uses, including cable insulation, heat shrinkable tubing and sheets, floppy discs, adhesive tape, car tyres and contact lenses.

The conventional method of curing surface coatings is a thermal process involving the use of organic solvents. About 20 million tons per year of these solvents are used world-wide in this process and about 40 per cent of it evaporates into the atmosphere to produce gases which contribute to the greenhouse effect. Electron beams can be used as an alternative method of curing which avoids this problem. At present, however, this technique accounts for only about 1 per cent of the market.

The environment

Tackling the problems of pollution and obtaining a better understanding of our global environment are vital to human development and in both these areas radiation techniques are making a significant contribution.

One such problem is the emission of flue gases into the atmosphere; electron beams are very efficient at removing the major harmful components – sulphur dioxide and the oxides of nitrogen. A pilot scheme has been built at a coal burning plant in Warsaw, Poland. This plant emits about 20 000 cubic metres of such gases per month and tests have shown that the device can remove 90 per cent of the sulphur dioxide and 86 per cent of the oxides of nitrogen. Three similar pilot schemes are operating in Japan. One is at a coal burning power station, the second at a municipal waste incinerator and the third removes traffic fumes in a tunnel.

Radiation, in the form of either an electron or gamma rays from cobalt-60, can be used to disinfect sewage sludge by destroying the harmful bacteria it contains. This is currently being done at two full scale plants, one in Germany and the other in India. The latter treats about 100 cubic metres of sludge per day. The irradiated sludge from the plants is used as organic fertiliser.

The chemical properties of scandium are such that it adheres to silt, so adding a small amount of scandium-46, a gamma emitter with a half-life of 84 days, to the silt on the bed of a river or estuary enables its movement to be traced. This technique has been used to study the movement of silt around the naval base at Rosyth in Scotland. For many decades the harbour had been dredged annually and the silt dumped out in the firth. One year it was decided to add some scandium-46 to the silt prior to dumping so that its subsequent movement could be determined. It was found that the silt was working its way back into the harbour; the same silt was being dredged out year after year, only to return in the intervening period. It was further discovered that if a site a few hundred metres away had been used for the dump then the currents were such as

Radionuclides are used to track the movement of pollution in rivers.

to take the silt out to sea. This new site was used following the next dredging and the silt has not been seen since. This discovery has been of considerable economic benefit to the Royal Navy, though less so to the dredging company.

An ability to date rocks can be helpful in mineral exploration. Many common rocks contain the natural radionuclides potassium-40, rubidium-87 and uranium-238. These have half-lives of 1.3, 4.0 and 4.5 billion years respectively, which are comparable with the age of many rock formations and hence are routinely used in dating procedures.

The accumulation of sediment on the seabed can be measured using thorium-230; studies have detected accumulation rates of the order of a few millimetres per thousand years. The movement of ocean currents can be traced using silicon-32 and argon-39 and atmospheric circulation can be followed with two radioisotopes of chlorine, 38 and 39. The rate of accumulation of ice on Antarctica has been measured using the radionuclide lead-210. The surprising result is that the continent turns out to be a cold desert with a water equivalent precipitation at the South Pole of only 6 cm per year.

Archaeology

All carbon in the environment contains carbon-14, a natural radionuclide with a half-life of 5 568 years, at a concentration of about a million millionth of a per cent. Living organisms continuously exchange carbon with the environment hence they also contain carbon-14 at this level. However, once the organism dies this exchange ceases and its carbon-14 level decreases at a rate determined by the half-life. So by measuring the amount of carbon-14 in an organic artefact, its age can be calculated.

A famous recent example of the application of this technique was the dating of the Turin Shroud. This is a relic reputed to be the burial sheet of Christ and hence arising from the first century AD. Alas the radiocarbon method indicated a date of 1325 AD, suggesting that it is a very clever medieval forgery which had foiled all other methods of dating. Another disappointment, for romantics, comes from the dating of the Round Table which hangs in Winchester Cathedral in

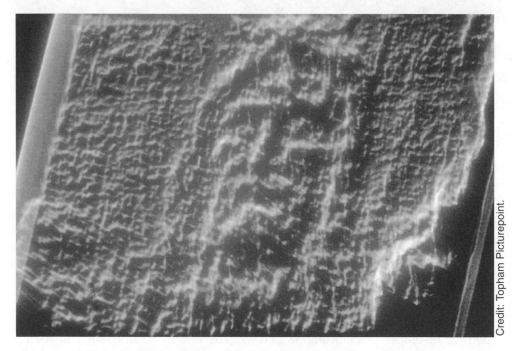

Carbon-14 dating was used to establish the age of the Turin Shroud.

England. This was believed by some to be the very one around which King Arthur sat with his Knights in the fifth century AD. Carbon dating, however, shows that it is a few centuries too young for that.

Natural radionuclides are present in the clay from which pottery was made in ancient times and can be used to date such artefacts through the phenomenon of thermoluminescence. Energy from radioactive decay is stored in the crystal structure of the minerals of the specimen and this is released in the form of light when it is subsequently heated, the amount of light being proportional to the time since the pot was fired.

Other areas

X-ray inspection units are widely used to check the contents of baggage or packages, for example at airports. The image is usually displayed on a television screen and can be electronically stored for subsequent examination.

Americium-241 is an actinide which finds application in the home. It is an alpha emitter with a half-life of 433 years and is used in domestic smoke alarms. Being electrically charged, the alpha particles are able to carry current between electrodes in the device, thus completing the circuit. They are easily stopped by smoke passing between the electrodes, which breaks the circuit and triggers the alarm.

The energy from radioactive decay ends up as heat, and this can be used as a small but vital source of power in situations where alternative sources would be inconvenient. The actinide plutonium-238, which is an alpha emitter with a half-life of 88 years, can be used for this purpose. It can be used to power space probes, particularly those on missions to the outer planets and beyond where solar panels cannot be used because of the huge distance from the sun. A Voyager space craft launched in 1977 is powered in this way. It reached the planet Neptune in 1989 and is now heading off into deep space. Its plutonium-238 powered battery is likely to cease functioning in the year 2007, by which time the Voyager will be 9 billion miles from earth. The same radionuclide has been used to power heart pacemakers.

Neutron activation analysis is an extremely powerful tool in forensic science – in the detection of arsenic for example, a poison which can be lethal at very low concentrations. Arsenic tends to concentrate in human hair and is normally present there at a level of about 0.8 part per million. If the concentration reaches a few hundred parts per million, the victim will almost certainly have received a lethal dose. When a human hair is irradiated in a reactor, the arsenic it contains will produce the radionuclide arsenic-76 which decays with a half-life of 26 hours to selenium-76, giving a characteristic set of gamma rays from which the concentration of arsenic can be determined down to the required level. Furthermore, human hair grows at a rate of about 0.3 mm per day. So analysis on small pieces of the same hair can determine the total amount of arsenic and the time before death at which it was administered – powerful forensic evidence.

Neutron activation analysis has been used to investigate the circumstances surrounding the death of Napoleon on St Helena in 1821. Arsenic was found in samples of his hair at levels far above normal. Whether it was administered by his British captors or by his doctor when treating the stomach cancer from which he reputedly died is still a matter of debate amongst historians.

Smoke detector, using americium-241.

Credit: Topham Picturepoint.

Chapter 3

SOURCES AND QUANTITIES OF RADIATION

People are exposed to radiation from many sources, both natural and man-made. The actual dose from any particular source can be measured or estimated. For most purposes, the unit of measurement is called the sievert (Sv); a useful sub-unit is the millisievert (mSv), one thousandth of a sievert.

World-wide, the average annual dose from natural sources is 2.4 mSv, although there are large national and regional variations around this figure. Of this, approximately 0.4 mSv is received from cosmic rays penetrating the earth's atmosphere. Approximately 1.3 mSv results from exposure to radon, a gas which seeps out of the ground; however, the dose related to radon exposure varies widely, depending on the local geology. The rest is received from various radioactive substances found naturally in our surroundings, our food, and in our own bodies.

The average dose from man-made sources varies depending on a number of factors including employment, medical history and lifestyle. The average dose to members of the public from medical applications of radiation and radioactive materials is 1 mSv a year in industrialised countries, with a global average of 0.3 mSv a year. The average annual dose from nuclear power generation is 0.001 mSv, rising to a few tenths of a mSv for a few individuals living close to a small number of nuclear installations. Particular human activities can result in an increased public exposure to natural sources, for example air travel, the use of phosphate fertilisers and the burning of coal. Workers in certain industries are exposed to higher than average doses; for example workers in mining, nuclear power generation and medical radiography, and aircrew.

Accidents can result in members of the public being exposed to much higher doses of radiation than normal. For example, some 250 000 people living around Chernobyl received dose of between 5 and 100 mSv in the year following the accident and some 9 000 received doses from 100 to over 200 mSv in that year. High doses have also resulted from accidents at military facilities and from loss or misuse of powerful medical sources.

People are exposed to radiation both from natural and from man-made sources. Most people get far more radiation from natural sources (often called background radiation) than from man-made sources. The exceptions include some patients who are being treated by radiation or who are undergoing some forms of medical investigation, and some of the people whose work involves the use of radiation or radioactive materials.

Some types of radiation exposure are more or less continual, some intermittent. Exposure to background radiation, being essentially inescapable, is continual, although it can vary, depending, for example, on whether one is indoors or in the open air. An X-ray investigation, on the other hand, usually involves an exposure of only about a second. Radiation exposure from radioactive material that is present or taken into the body continues until it is excreted, or until its radioactivity has died away, a time depending on its half-life.

HALF-LIFE

Radioactivity decreases with time as the radionuclides decay into stable atoms. The rate of decay is governed by the half-life of the radionuclide. After one half-life, the radioactivity falls to half its original value, after two to one quarter, after three to one eighth, and so on. Half-lives vary from fractions of a second to millions of years.

| After one half-life | After two half-lives | After three half-lives | After four half-lives | After five half-lives | After six half-lives | After seven half-lives |

In order to compare radiation exposures from different sources, appropriate units are needed. The measure that is used for quantifying radiation exposure is dose. A dose of medicine is usually expressed as the weight of the medicine swallowed, injected or inhaled. In the case of radiation, the dose depends on the energy absorbed by the body from the radiation, since it is the absorption of this energy that initiates the physical and chemical changes that may lead to

32

biological harm. The amount of harm that results depends not only on the amount of energy absorbed, but also on the type of radiation and on the part of the body that is irradiated. The units used to measure radiation dose take all these factors into account (see box on "Units" on pages 34-35).

HIERARCHY OF UNITS

ABSORBED DOSE [gray (Gy)]
energy imparted by radiation to unit mass of tissue

EQUIVALENT DOSE [sievert (Sv)]
absorbed dose weighted for biological effectiveness
of different types of radiation

EFFECTIVE DOSE [sievert (Sv)]
equivalent dose weighted for radiosensitivity of
different tissues

COLLECTIVE DOSE [person-sievert (person-Sv)]
average effective dose times the number of
people exposed

Natural sources

Mankind has evolved in a naturally radioactive environment. The earth is bombarded by cosmic rays from space, and all matter contains some traces of radioactive substances. People are exposed to external radiation, consisting of cosmic radiation and radiation from the naturally occurring radionuclides in their immediate surroundings, and to internal radiation, from the naturally occurring radionuclides that are taken into their bodies in the form of food, drink and air. The average annual dose from all these sources combined is around 2.4 mSv, with large variations around that figure.

33

UNITS

The amount of energy that is absorbed from the radiation by a unit mass of the matter through which it passes is called the **absorbed dose**. Thus if 1 Joule (J) of energy is absorbed by 1 kilogram (kg) of tissue, the absorbed dose is 1 J/kg, or 1 gray (Gy), named after a pioneer researcher in radiation measurement and radiobiology.

Different types of radiation interact with matter in different ways. For example, an alpha particle, which is relatively heavy on the atomic scale, will cause a series of closely spaced ionisations, whereas the ionisations caused by a far lighter beta particle (an electron), or a gamma ray will be much more spaced out. The biological consequences of these two different ionisation patterns are distinctly different. Since in general it is the biological effectiveness of a dose of radiation that is of interest rather than the physical amount of energy absorbed, a further unit has been introduced, which is the dose in grays multiplied by a weighting factor which takes into account the different biological effectiveness of different types of radiation. The resulting unit is called **equivalent dose**.

Furthermore, different organs and tissues will be affected by radiation in different ways. Their different radiosensitivities are taken into account by multiplying the equivalent dose by a further weighting factor, depending on the organ or tissue irradiated. The resulting unit is then called **effective dose**.

Equivalent and effective doses are expressed in J/kg, but the name sievert (Sv), after an eminent Swedish radiation physicist, is used to avoid confusion with absorbed dose (Gy). The sievert is an inconveniently large unit for the sort of radiation doses normally received, for example from one year's exposure to natural background radiation, so the millisievert (mSv), which is one thousandth of a sievert, is often used instead.

The weighting factors used to derive equivalent and effective dose in sieverts from absorbed dose in grays apply only to the stochastic effects of radiation, cancers and hereditary effects; the sievert is the unit used when discussing such effects. In the remainder of this report, when a dose is expressed in mSv it refers to effective dose and relates to stochastic effects (stochastic and deterministic effects are explained in Chapter 4).

Absorbed dose, equivalent dose and effective dose are commonly shortened to "dose", the unit referred to generally being clear from the context.

The total impact of the radiation exposures that result from a given source or activity depends on the number of people exposed as well as on the doses they receive. The measure used is **collective dose**, which is the average (effective) dose times the number of people exposed. The unit of collective dose is the person-sievert (person-Sv). Although the concept of collective dose may be useful for making comparisons between different activities that involve radiation exposure, it is not used for setting regulatory limits, which are discussed in Chapter 5.

The becquerel (Bq) is the unit used to measure radioactivity, named after its discoverer. A quantity of radioactive material that has a radio-activity (often abbreviated to activity) of 1 becquerel is one in which one unstable nucleus changes and emits radiation every second. The becquerel is a very small unit, for example an adult human contains several thousand becquerels of the naturally radioactive form of potassium; the kilobecquerel (kBq or thousand Bq), the megabecquerel (MBq or million Bq) and the gigabecquerel (GBq or thousand million Bq) are commonly used.

Cosmic radiation

The primary cosmic radiation incident on the top of the earth's atmosphere comes mainly from beyond the solar system, some probably even from beyond our galaxy, with a small component from the sun. It consists of radiation with a wide range of energies, mainly protons, but also alpha particles and some heavier nuclei, electrons, and gamma rays. The incoming flow of cosmic radiation is more or less constant and uniform, but the lower energy components are modified by the earth's magnetic field, resulting in somewhat higher intensities in polar regions, and by solar activity, resulting in small variations with the eleven year solar cycle.

The primary cosmic radiation is substantially altered by its passage through the atmosphere, with much of it being absorbed before it reaches sea level. As a result the main factor influencing the doses received by people from cosmic sources is the altitude at which they live.

The average dose that people get from cosmic radiation is about 0.4 mSv a year. People living at about 1 000 metres, the height of a typical ski resort, get about 20 per cent more cosmic radiation than people living at sea level. Air travel exposes people to much higher dose rates. The dose rate at the cruising height of a jet aeroplane, for example, is over 150 times that at sea level, and it is even higher for a supersonic plane.

Naturally occurring radionuclides

Some radionuclides are produced as a result of the passage of cosmic rays through the atmosphere, but these result in very small doses to people. The largest contribution to background radiation comes from radionuclides whose half-lives are of the same order of magnitude as the age of the earth itself – thousands of millions of years. They are the remnants of the primordial inventory of radionuclides, and the members of their decay series, also known as their radioactive daughters. For example, the most common form of uranium, uranium-238, decays through a series of radionuclides which includes radium and radon (a radioactive gas). The end product of this decay chain is stable lead. Another important naturally occurring radionuclide is a radioactive form of the common element potassium (potassium-40), with a half-life of 1 260 million years.

The external doses received from natural radionuclides in the soils and rocks depend on the local geology and factors such as moisture content, snow cover and atmospheric conditions. Indoors, doses depend on the extent of shielding from the outdoor exposure provided by the building itself, and on the construction materials used. The average dose that people get from this source is about 0.5 mSv a year, but there are large variations around this average, with many people getting ten times the average and a few communities, living near some types of mineral sand, getting up to one hundred times the average.

The internal doses received from natural radionuclides come almost entirely from potassium-40, and from radon and thoron and their decay products. Potassium is an essential ingredient of body cells, and an average adult man contains about a tenth of a kilogramme, of which about 16 milligrammes is potassium-40. The average dose received from this source is about 0.2 mSv a year, and varies little from person to person.

Radon and its decay products (and, of somewhat less importance, thoron and its decay products) are the largest single sources of radiation exposure for most people. The radon and thoron come from the decay of uranium and thorium in the earth's crust. These gases seep out of the ground, at a rate that depends on geology, soil condition and cover, etc. In the open air, they are quickly dispersed and their concentrations, and hence the resulting doses when they are inhaled, are low. However, when they enter a building, for example by seeping through the floor, through cracks or past water and drainage pipes, or are emitted from the natural radionuclides contained in building materials, concentrations build up unless the building is very well ventilated. The gases themselves are chemically inert and only slightly radioactive, and give very small direct doses. However their radioactive decay products (mainly polonium, bismuth and lead radionuclides) are reactive, and consequently attach to dust particles and water droplets. These may be breathed in and settle on the surfaces of the lung, which are irradiated as a result. A very wide range of doses results from this source, depending on the local geology and on building materials, construction methods and ventilation. The average dose is 1.3 mSv a year, but exposures range up to a hundred times the average, and in some rare and extreme cases, for example some homes built over old mine workings, over one thousand times the average.

Natural sources and human activity

In addition to the various sources of natural radiation described above, there are several ways in which human activities can result in enhanced exposures to naturally occurring radionuclides. Two of these are coal mining, and the mining of phosphate rock for use in fertilisers and other products. Coal and phosphate rock both contain traces of naturally radioactive materials such as uranium and radium. Although the concentrations are low, the quantities extracted are very large and the materials become widely distributed in the environment, either directly as phosphate fertiliser or indirectly as a result of the coal being burned and a small proportion of the ash being emitted through the chimney. The individual doses received are very small, but they can result in large collective doses because of the very large numbers of people exposed. The annual global collective dose from the phosphate industry, for example, is over ten times that from routine operations of the whole of the nuclear industry.

Pathways for radioactive material through the environment that may result in radiation exposures

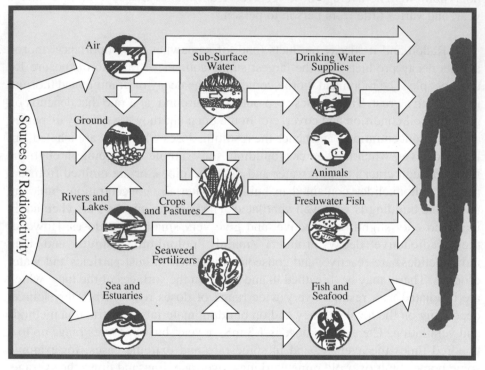

Man-made sources

The uses of radiation and radioactive materials have expanded greatly, particularly since the discovery and development of nuclear fission and the availability of a wide range of man-made radionuclides. Few people in OECD countries have not benefited in some way, either directly, as a result of medical uses, or indirectly, in energy production, manufacturing industry, agriculture and pollution control; all these result in radiation exposures.

Medical exposures

Medicine is the largest source of man-made radiation exposure in most countries. Diagnostic X-rays are the commonest form, and in most industrialised countries there are about 900 medical (*i.e.* non-dental) X-ray examinations per thousand people per year. Typical doses range from 0.1 mSv for a chest X-ray to 5-10 mSv for an upper gastrointestinal tract examination or an angiograph. The resulting average dose is about 1 mSv a year in industrialised countries, with a world average of about 0.3 mSv a year.

The use of radioactive materials to study body processes and locate tumours has increased rapidly over the past 30 years, but these techniques are still far less used than X-rays. Radiation doses are typically between 1 and 20 mSv. The average annual doses received, both for industrialised countries and for the world population are about one tenth of those for X-rays.

By far the largest doses in medicine are those used in treating cancer, but the number of patients receiving such treatment is far lower than the number exposed to radiation for diagnostic purposes. A typical treatment would involve doses of a few tens of grays, usually given in a number of separate irradiations. The radiation is carefully focused on the tumour itself to minimise damage to surrounding tissues.

Radiation exposures from these medical practices are received mainly by the patients themselves, but there is also inevitably some exposure of doctors, nurses, radiographers, other hospital workers and people disposing of any radioactive sources used.

Fallout from nuclear explosions

Nuclear weapons tests during the 1950s and early 1960s resulted in radioactive materials being spread throughout the atmosphere. Most of this material has now been deposited on the land and the oceans. All weapons testing since 1980 (and most since the Test Ban Treaty of 1963) has been carried out underground; this gives rise to virtually no fallout. The average annual dose from nuclear weapons tests peaked at about 0.1 mSv in 1963-64; this has since fallen to less than one-twentieth of the peak level and, provided that current treaties continue to be effective in ensuring that no further atmospheric testing takes place, doses will continue to fall.

Annual radiation doses from nuclear weapons testing

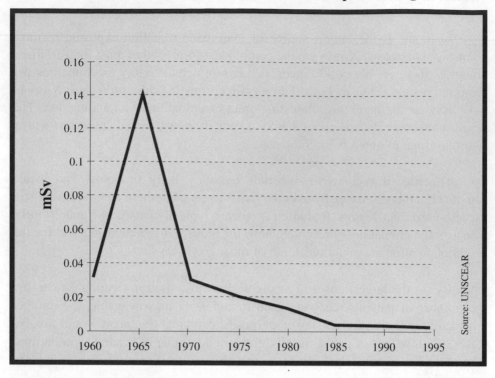

Source: UNSCEAR

Nuclear power production

The generation of electricity in nuclear power stations involves the mining and processing of uranium to fuel the reactors, the operation of the reactors themselves, and the transport, treatment and disposal of the radioactive waste products. In some countries, spent fuel is reprocessed to separate uranium and plutonium from the waste products for reuse; other countries plan to dispose of spent fuel without reprocessing. All these activities result in very small releases of radioactive materials to the environment and very small radiation doses to members of the public, ranging from virtually nothing for people living far from any nuclear installation to a few tenths of a mSv a year for a handful of people living near a very few installations; the average is about 0.001 mSv a year.

Major accidents

The normal operations of nuclear power stations generally contribute an extremely small fraction of the total radiation exposure of people living nearby. Much larger doses can occur as a result of major accidents. The two most serious accidents to nuclear power stations were at Three Mile Island in the United States, in 1979, and Chernobyl in the Soviet Union, in 1986.

The Three Mile Island accident caused serious damage to the reactor core, but the built-in protective features of the reactor ensured that very little radioactivity was released to the environment. Three workers received doses of between 30 and 40 mSv; doses to members of the public were low, the highest being just below 1 mSv.

The Chernobyl accident in 1986 resulted in the release of a large fraction of the radioactivity in the reactor core. Many workers were exposed to high doses, particularly during the heroic efforts to bring the burning reactor under control and the immediate clean-up operations. A total of 31 workers died as a result of the accident and about 140 suffered various degrees of radiation sickness and health impairment. No members of the public suffered these kinds of effects, but more than 100 000 people were evacuated, mostly from the 30 km radius around the accident. These people received an average of about 15 mSv before they were evacuated, and some continued to be exposed thereafter, although to a lesser extent, depending on the sites of their relocation. About 270 000 people living in contaminated areas of the former Soviet Union received an average of 40 mSv for the 1986-89 time period. Hundreds of thousands of workers, including many military personnel, were involved in the emergency actions and subsequent clean-up operations; many received significant doses, ranging from tens to hundreds of millisieverts. The possible health consequences to people in the contaminated regions of the former Soviet Union are discussed in the next chapter. Some volatile radionuclides from the accident, such as iodine and caesium, spread throughout the Northern hemisphere; doses depended mainly on whether it was raining while the radioactive cloud was passing. Doses ranged from a few microsieverts (millionths of a sievert) outside Europe to an upper extreme of 1 or 2 mSv in some European countries, which is similar to annual individual exposures from natural background radiation. The estimated total global collective dose that can be ascribed to the accident is 600 000 person-Sv, 53 per cent of this being received in European countries and 36 per cent in the Soviet Union. About one third of this collective dose was received within one year of the accident, the remainder being spread over some tens of years.

There have been a number of accidents related to nuclear weapons production and transport. The two most serious both occurred in 1957, one at Kyshtym in the Soviet Union and one at Windscale in the United Kingdom. At Kyshtym a failure of the cooling system in a storage tank containing highly radioactive waste led to a chemical explosion. An area of 15 000 square kilometres, with 270 000 inhabitants, was contaminated. Some 10 000 people were evacuated, a measure that may have reduced exposures by a factor of ten. Those not evacuated received a dose estimated at 12 mSv, spread over a period of 30 years. At Windscale, a reactor designed to produce plutonium for weapons caught fire. The resulting doses, even to the most exposed members of the public, were below 1 mSv; the highest worker dose was 47 mSv.

Other accidents, mostly during the early days of nuclear weapons research and development, are known to have resulted in deaths of 8 workers, but there is a lack of reliable data from such activities in the Soviet Union and China and the total could well be higher. Another example of extremely high worker doses, associated not with an accident but with "routine" operations, has only recently come to light. Soviet mining for uranium in Germany during the immediate post-war period resulted in mean annual doses for large numbers of workers corresponding to a staggering 750 mSv (effective dose), which corresponds to about 6 Gy per year to the lung. There are fears that the final toll of lung cancer from uranium mining in Saxony and Thuringia during that period may reach as high as 20 000. There is also a legacy of several kilogrammes of radium, along with other pollutants, in the surrounding environment, now the subject of a major clean-up operation.

The loss, damage or misuse of powerful sources and devices used for industrial or medical applications, including a few cases of medical "maladministration", has sometimes led to serious exposures, with a total of 44 fatalities since 1960, including several among members of the public. The most serious events were in Mexico in 1962 (4 deaths), Morocco in 1984 (8 deaths), Goiana, Brazil, in 1987 (4 deaths) and Spain, in 1990 (11 deaths).

Miscellaneous sources

Most people are exposed to radiation from a variety of man-made sources. These include watches and clocks luminised with radioactive material, although production of these stopped many years ago, and smoke detectors. However the resulting doses are small; such sources, together with the exposures to natural radioactive materials released as a result of human activities (fertiliser and coal ash) and extra cosmic radiation received during air travel, contribute on average less than 0.001 mSv a year to average radiation exposures.

routine activities

the largest doses of radiation from man-mad
work in the nuclear power industry or with radi
medicine, research and industry. World-wide, a
sed to man-made radiation as a result of their wo
ween different occupations and from country to
es, based on comprehensive dose monitoring syst
eactor operators at nuclear power stations, 1 mSv a
phers and 0.5 mSv a year for medical radiation work

kers are exposed to higher than average doses from th
r example aircrew, who get typically 2 to 3 mSv a year
ls of cosmic radiation at high altitudes, and miners of
ls, who get typically 1 to 10 mSv a year as a result of e
vels in many mines. Workers at radon spas, caves in which t
entrations of radon, where many people go in the belief tha
their health, can get appreciably higher doses. These estima
n because the radiation exposures of most of these workers
ly monitored.

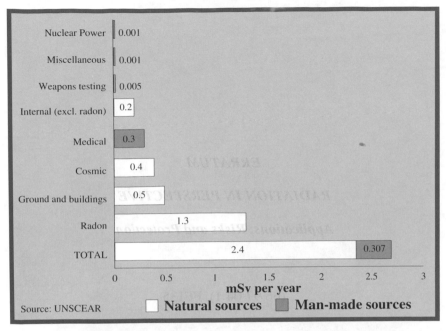

Source: UNSCEAR

Natural sources **Man-made sources**

Nuclear industry: annual collective occupational doses

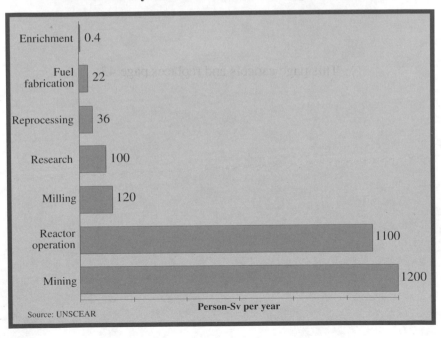

Source: UNSCEAR

43

44

ERRATUM

RADIATION IN PERSPECTIVE

Applications, Risks and Protection

(66 97 04 1) FF135

This page cancels and replaces page 43.

Annual radiatio

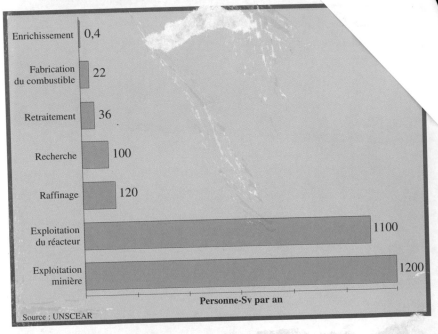

Production électronucléaire	0,00
Divers	
Essais	
Interne l'exclusion du rado	
Professions médicales	
Rayonnement cosmique	0,4
Sols et bâtiments	0,5
Radon	1,3
TOTAL	

0   0,5   1   mSv

Source : UNSCEAR   ☐ **Sources naturelles**

Occupational exposures fro

The people who get
are generally those who
radioactive materials in
million people are expo
vary considerably be
Typical average dos
2.5 mSv a year for
industrial radiogra

Some wo
background, f
increased lev
other miner
radiation le
high con
improve
uncerta
regula

Nuclear industry: annual collective occupa

Enrichissement	0,4
Fabrication du combustible	22
Retraitement	36
Recherche	100
Raffinage	120
Exploitation du réacteur	1100
Exploitation minière	1200

Personne-Sv par an

Source : UNSCEAR

Annual radiation doses from all sources
(world average)

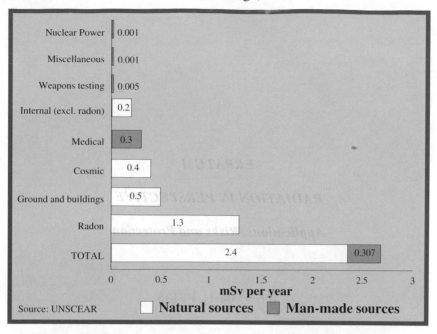

Source: UNSCEAR

☐ **Natural sources** ▨ **Man-made sources**

Nuclear industry: annual collective occupational doses

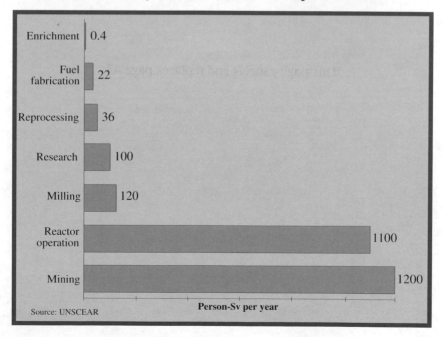

Source: UNSCEAR

ERRATUM

RADIATION IN PERSPECTIVE

Applications, Risks and Protection

(66 97 04 1) FF135

This page cancels and replaces page 43.

Annual radiation doses from ~~1~~ sources
(world average)

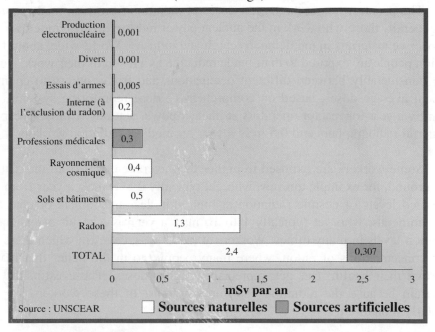

Production électronucléaire — 0,001
Divers — 0,001
Essais d'armes — 0,005
Interne (à l'exclusion du radon) — 0,2
Professions médicales — 0,3
Rayonnement cosmique — 0,4
Sols et bâtiments — 0,5
Radon — 1,3
TOTAL — 2,4 | 0,307

0   0,5   1   1,5   2   2,5   3
mSv par an

Source : UNSCEAR   □ **Sources naturelles**   ■ **Sources artificielles**

Nuclear industry: annual collective occupational doses

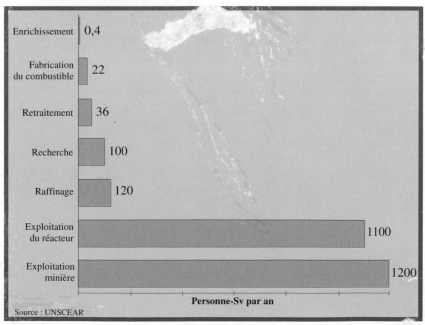

Enrichissement — 0,4
Fabrication du combustible — 22
Retraitement — 36
Recherche — 100
Raffinage — 120
Exploitation du réacteur — 1100
Exploitation minière — 1200

Personne-Sv par an

Source : UNSCEAR

Occupational exposures from routine activities

The people who get the largest doses of radiation from man-made sources are generally those who work in the nuclear power industry or with radiation and radioactive materials in medicine, research and industry. World-wide, about four million people are exposed to man-made radiation as a result of their work. Doses vary considerably between different occupations and from country to country. Typical average doses, based on comprehensive dose monitoring systems, are 2.5 mSv a year for reactor operators at nuclear power stations, 1 mSv a year for industrial radiographers and 0.5 mSv a year for medical radiation workers.

Some workers are exposed to higher than average doses from the natural background, for example aircrew, who get typically 2 to 3 mSv a year from the increased levels of cosmic radiation at high altitudes, and miners of coal and other minerals, who get typically 1 to 10 mSv a year as a result of enhanced radiation levels in many mines. Workers at radon spas, caves in which there are high concentrations of radon, where many people go in the belief that it will improve their health, can get appreciably higher doses. These estimates are uncertain because the radiation exposures of most of these workers are not regularly monitored.

COMPARISONS

The United Nations Scientific Committee on the Effects of Atomic Radiation (UNSCEAR), the prime source of authoritative information on sources and effects of radiation, has recently introduced a method of presenting information on radiation exposures which provides a useful perspective on the implications of the doses received from man-made sources compared with those from natural sources.

"On a global basis, one year of medical practice at the present rate is equivalent to about 90 days of exposure to natural sources, but individual doses vary from zero (for persons not examined or treated) to many thousands of times that received annually from natural sources (for patients undergoing radiotherapy).

Most of the doses committed by one year of current operations of the nuclear fuel cycle are widely distributed and correspond to about 1 day of exposure to natural sources. Excluding severe accidents, the doses to the most highly exposed individuals do not exceed, and rarely approach, doses from natural sources.

Occupational exposure, viewed globally, corresponds to about 8 hours of exposure to natural sources. However, occupational exposure is confined to a small proportion of those who work. For this limited group, the exposures are similar to those from natural sources. For small sub-groups, occupational exposures are about five times those from natural sources.

The collective dose committed (by atmospheric nuclear testing) corresponds to about 2.3 years exposure to natural sources. This figure represents the whole programme of tests and is not comparable with the figures for a single year of practice.

Only one accident in a civil nuclear power installation, that at Chernobyl, has resulted in doses to members of the public greater than those resulting from the exposure in one year to natural sources. On a global basis, this accident corresponded to about 20 days exposure from natural sources."

COLLECTIVE DOSE COMMITTED TO THE WORLD POPULATION BY A 50-YEAR PERIOD OF OPERATION FOR CONTINUING PRACTICES OR BY SINGLE EVENTS FROM 1945 TO 1992

SOURCE	BASIS OF COMMITMENT	COLLECTIVE EFFECTIVE DOSE (million person-Sv)
Natural sources	Current rate for 50 years	650
Medical exposure	Current rate for 50 years	
Diagnosis		90
Treatment		75
Atmospheric nuclear weapons testing	Completed practice	30
Nuclear power	Total practice to date	0.4
	Current rate for 50 years	2
Severe accidents	Events to date	0.6
Occupational exposures	Current rate for 50 years	
Medical		0.05
Nuclear power		0.12
Industrial uses		0.03
Defence activities		0.01
Non-uranium mining		0.4
Total (all occupations)		**0.6**

Source: UNSCEAR 1993

Chapter 4

EFFECTS OF RADIATION

Radiation can affect the body by damaging or destroying cells. High doses of radiation can result in the destruction of many cells, which in turn can result in minor skin irritation, burns or other serious damage, or even death, depending on the size of the dose. Such effects are known as deterministic, because they can be directly associated with a particular exposure to radiation. Such effects are not usually apparent at doses of less than 1 Gy.

If the deoxyribonucleic acid (DNA) of a cell is damaged, it is possible that subsequent divisions of that cell will produce abnormalities, which can develop into cancer. Damage caused to cells in the ovaries or testes can result in abnormalities that will only become apparent in the offspring of the exposed individual. The higher the dose, the more likely it is that cells will be damaged in such a way as to result in cancer or inherited abnormalities. Such effects are called stochastic effects because it is the probability rather than the severity of harm which increases with the size of the dose.

While deterministic effects are relatively easy to measure by direct observations of affected individuals, stochastic effects can only be measured by statistical analysis of a large exposed population and comparison with a similar non-exposed population. For example, the number of cancers that can be ascribed to radiation from the Hiroshima and Nagasaki bombings has been established by studying the long term health of the victims and their children. Such analysis is complicated by the fact that the "normal" incidence of cancer in any given population is already high, typically around 25 per cent. Further evidence is available from the long-term medical records of several groups of patients exposed to radiation for diagnostic or therapeutic purposes, some 40 000 people in total, and from records of groups of workers who have had higher than average occupational exposure.

Extensive studies of stochastic effects have resulted in conclusive evidence for an enhanced risk of cancer associated with doses over about 50 mSv. There is no reliable evidence of cancer risk from lower doses, and no inherited damage has been found in humans at any dose level. Estimates can be made concerning the effects of lower doses by extrapolating the results obtained from studies of populations exposed to higher doses. For inherited damage, some estimates can be made on the basis of animal studies. These estimates of the risks of cancer and inherited damage, together with an overall assumption that no exposure is entirely risk free, form the basis of international recommendations on radiation protection.

The miracle of life is possible because the cells that are present in all living matter are able to grow and reproduce themselves. Radiation is one of the many ways in which these processes can be affected. Radiation can be thought of as a carrier of energy, kinetic energy in the case of alpha and beta particles, and electromagnetic energy in the case of X-rays and gamma rays. The absorption of some of this energy by the atoms or molecules that make up the cells can result in one or more ionisations, which rapidly lead to a series of chemical changes. All molecules can be altered by radiation in this way; in living matter the most important targets are the DNA molecules. These are the crucial constituent of chromosomes which carry the genetic information and other cell regulatory processes from a cell to the next generation when it divides.

Ionisation and subsequent changes to DNA occur continuously as a result of the interaction of natural radiation with the body. The average person, for example, is exposed to over 200 million gamma rays from external sources every hour throughout life, in addition to the radiation from the decay of about 15 million potassium-40 atoms in the body, again occurring every hour throughout life. The absence of any evidence of harm from such exposures suggests that most of the damage caused by such irradiation is either unimportant or readily repaired by the body's natural repair mechanisms. There are, however, two ways in which the damage can be important:

- the damage to the DNA is not repaired, or is repaired incorrectly, and the cell which contains that DNA dies, or its subsequent division results in non-viable daughters;

- the damaged cell is able to divide and produce viable daughters, but these carry the uncorrected errors in the DNA and function abnormally as a result. The abnormality may affect the subsequent life of the organism that has been irradiated, generally in the form of cancer. If the damage occurs in a cell in the testes or ovaries which is involved in the conception of a new individual, the error can lead to inherited abnormalities in the next generation and possibly in subsequent generations.

48

Deterministic effects

An adult person contains about 60 million million cells. Several million of these die every day and are replaced by new ones. The cells that are killed by a low or moderate dose of radiation are replaced within a few days or weeks, with no effect on the way the body or the part that has been irradiated functions. However, if the number of cells killed is large enough, the body may not be able to replace them quickly enough. The effects range from relatively trivial (a slight reddening of the skin), to very serious (major burns and other injuries). At high enough doses, death occurs within days or weeks. These effects are called deterministic, or early (since they appear within a short time of the radiation exposure). They are certain to occur above a certain level of dose, called the threshold, and above this level their severity increases with dose.

The threshold for deterministic effects depends on the tissue being irradiated. A typical value is around 1 Gy, corresponding to a dose some 500 times greater than the average annual dose from background radiation and a million times higher than average public doses from nuclear power production. The threshold for some effects (temporary sterility in the male) is up to a factor of five lower; that for others (permanent sterility, opacity of the lens of the eye) about a factor five higher. Survival is unlikely following doses above around 10 Gy to the whole body. Doses above the threshold for deterministic effects have only occurred as a result of nuclear weapons explosions and a small number of serious accidents, including Chernobyl. The doses have to be delivered in a relatively short time, typically less than a few hours, if they are to produce any harmful effect. If the same dose is spread out over a longer period, the body may be able to replace the dead cells and not show any effects.

While the threshold for deterministic effects is well above the exposure levels experienced in normal circumstances, there is one important exception: exposure of the developing embryo. During early development, there are certain stages at which only a few cells may be responsible for the formation of whole organs or parts of the body. Even a small amount of cell killing at these stages can result in serious malformations, and the threshold for such effects is estimated to be around 0.1 Gy. At a somewhat later stage of pregnancy, at between 8 and 15 weeks, the destruction of a small number of cells in critical organs such as the brain or the nervous system can still have severe consequences, leading, for example, to serious mental retardation. Particular care therefore has to be taken to avoid unnecessary radiation exposures during pregnancy.

Deterministic effects can be highly beneficial: they are the basis of radiotherapy, in which large doses of radiation are used to destroy cancerous cells. The radiation can come from an external source, carefully focused to minimise damage to surrounding tissues, or it can be produced internally, from the decay of a radionuclide which is concentrated selectively in the part of the body being treated.

DETERMINISTIC AND STOCHASTIC EFFECTS

DETERMINISTIC (EARLY) EFFECTS

- certain to occur above a threshold level of dose
- severity increases with dose
- threshold ≈1 Gy (lower for the developing embryo)
- survival unlikely above ≈10 Gy (whole body)

STOCHASTIC (LATE) EFFECTS

- probability of occurrence (rather than severity of effect) increases with dose
- evidence of risk of cancer for doses above ≈50 mSv
- no evidence below ≈50 mSv, but assumption that risk is proportional to dose; no evidence of hereditary defects at any dose level in humans
- assumption that risk of hereditary defects is proportional to dose (from animal studies)

Stochastic effects

Damage which results in viable but altered DNA has markedly different consequences from damage which results in cell killing. Cell killing will cause harm at doses above the threshold level, with a higher dose leading to more severe harm. A viable alteration in DNA may cause cancer, or an inherited abnormality in one or more future generations, and it is the probability of such an outcome occurring, not the severity, that depends on the amount of radiation received. Any cancer that can be caused by radiation may result from either a small or a large dose, but it is more likely to result from a large one. In the same way, an inherited abnormality may result from either a small or a large dose, but is more likely to result from a large one. Since it is the probability and not the severity of the consequence that depends on the dose, the effects are called stochastic – meaning of a random or statistical nature, or late or delayed (since they do not appear until after a latent period ranging from a few years to several decades).

Unlike deterministic effects, it is not generally possible to associate a particular cancer or inherited abnormality in an individual with a particular radiation exposure or history of such exposures. There are many possible causes of cancer and inherited abnormalities, and there are large variations in the incidence of such conditions. If a person is exposed to a dose of radiation and later develops cancer, it does not follow that the exposure caused the cancer. Indeed, radiation-induced cancer is a relatively rare phenomenon. For example, 7 827 deaths from cancer occurred among the 86 572 survivors of the Hiroshima and Nagasaki bombings between 1950 and 1990 (7 578 "solid" cancers and 249 leukaemias). However, only 421 of these (334 "solid" cancers and 87 leukaemias) can be ascribed to radiation from the bombs. The incidence both of "natural" and of radiation-induced cancers in this population is expected to grow as the population ages. However, leukaemia peaks about 7 years after exposure for children and a little later in adults; the radiation-induced leukaemias at Hiroshima and Nagasaki peaked around 1955.

Relationships between stochastic effects and radiation exposures can only be detected by statistical studies of large populations, that is by epidemiology. A major problem is that the "normal" incidence of lethal cancer is around 25 per cent, and total incidence for all cancers may reach 30 per cent, with substantial variations around these figures. The expected increase in cancer incidence resulting even from a major accident such as Chernobyl is very small in comparison with these natural rates and study of very large samples of populations is needed to give statistically significant results. Useful additional information on the mechanisms of cellular damage and on the consequent effects of such damage on tissues can be gained from studies in the laboratory, but such studies cannot be used to derive numerical values of the risks of radiation to humans.

Evidence for radiation-induced cancer

While the first indications of the damage that could be caused by radiation were of deterministic effects, particularly skin burns, it was soon realised that Lister's warning "that the transmission of radiation through the body was not altogether a matter of indifference to the internal organs" was justified. Early radiologists took few precautions to limit their exposures and began to experience higher than expected leukaemia and skin cancer rates. Excess cancers were also found among some patients with non-malignant diseases which were treated by radiation. Since there were essentially no quantitative means of measuring doses, however, it was not possible to derive quantitative risk estimates from these early observations.

Possible consequences of radiation exposure

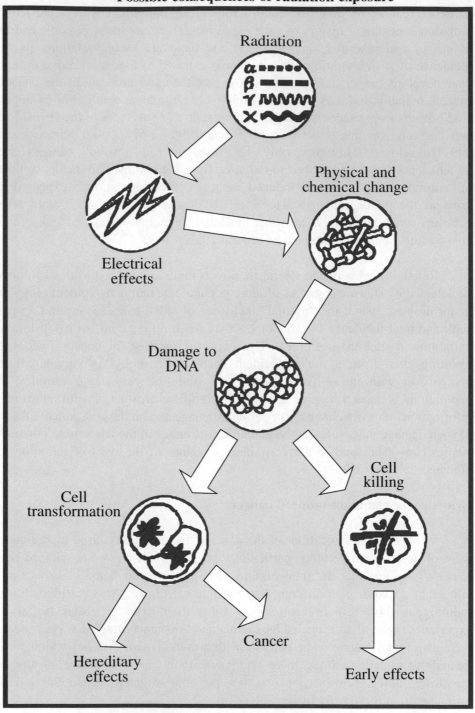

There is now a broad set of studies on which quantitative estimates of the risk of cancer induction by radiation can be based. The evidence comes from three classes of radiation exposure: military, medical and occupational. In all the studies, the doses that were associated with enhanced cancer incidence were of the order of 50 mSv or above. The evidence is based mainly on a total of around 2 000 cancer deaths world-wide, including those at Hiroshima and Nagasaki, that can definitely be ascribed to radiation and for which reasonably accurate estimates of dose can be made.

The most important source of information is the studies of the Hiroshima and Nagasaki survivors, in terms of numbers of people exposed, collective dose received, follow-up period and excess fatal and non-fatal cancers that have occurred. These studies have provided the most detailed information available on the risks of induction of different types of cancer by radiation, for doses above about 50 mSv. The exposures from the bombs were very intense and brief; it cannot be assumed that the risks derived from these studies can be applied directly to different exposure conditions. Up to 1990 the cancer rate among the survivors was nearly 6 per cent higher than expected in a similar Japanese population. This figure might reach around 10 per cent during the remaining life span of the survivors, more than half of whom are still alive – cancer incidence generally increases with age. The additional information on cancer incidence in the ageing population that will become available during the next few decades will give greater confidence in the risk estimates.

There has been a real and significant excess of thyroid cancers among children living in the contaminated regions around Chernobyl in the decade following the accident. This excess should be ascribed to the accident until proved otherwise, and it may continue for some time. Large epidemiological programmes are in place, some sponsored by international organisations such as the World Health Organisation. However, on the basis of generally accepted estimates of the doses received, it is unlikely that any discernible radiation effects in the general population above the background of natural incidence of the same diseases will be found, other than thyroid disease. No increase in any other kind of cancer or leukaemia, congenital abnormalities, adverse pregnancy outcomes or any other radiation-induced disease that could be ascribed to the accident have been observed. There has, however, been a significant amount of psychological stress and consequent health problems in the populations affected, mainly in the contaminated regions of the former Soviet Union.

Most of the medical evidence comes from about ten groups of patients exposed to a range of treatments and diagnoses with radiation and radioactive

Effects of radiation

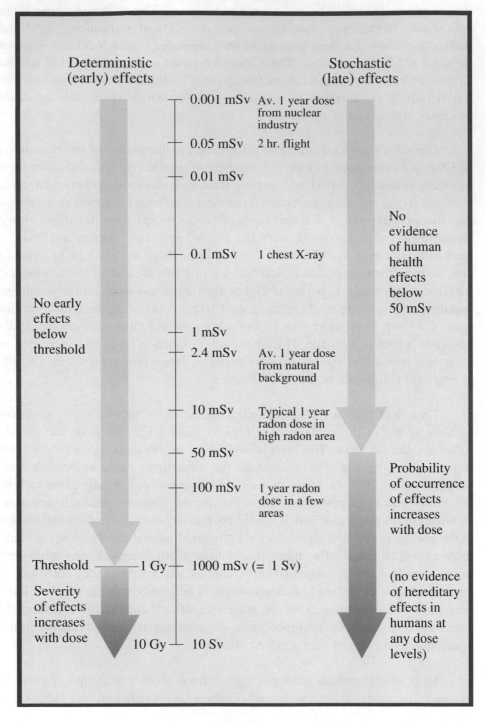

Deterministic (early) effects

Stochastic (late) effects

0.001 mSv	Av. 1 year dose from nuclear industry
0.05 mSv	2 hr. flight
0.01 mSv	
0.1 mSv	1 chest X-ray
1 mSv	
2.4 mSv	Av. 1 year dose from natural background
10 mSv	Typical 1 year radon dose in high radon area
50 mSv	
100 mSv	1 year radon dose in a few areas

No early effects below threshold

No evidence of human health effects below 50 mSv

Threshold —— 1 Gy — 1000 mSv (= 1 Sv)

Severity of effects increases with dose

10 Gy — 10 Sv

Probability of occurrence of effects increases with dose

(no evidence of hereditary effects in humans at any dose levels) materials – about 40 000 patients in all. Much of the information comes from exposures that occurred many decades ago, and there is some uncertainty in the dose figures. However, the availability of medical records giving details of the treatment and the subsequent health of the patients, as well as information about patients with similar illnesses who were not treated with radiation, gives reasonable confidence in the risk estimates derived for the exposure conditions used and the cancers involved. With today's knowledge of the risks of radiation, and the availability of better techniques that allow smaller doses to be used and alternative treatment methods for some conditions previously treated with radiation, it is unlikely that a large amount of additional data will be forthcoming. Nevertheless, it is likely that there will continue to be many medical circumstances when the small long-term risk of cancer being caused by a treatment or diagnosis by radiation will be greatly outweighed by the shorter-term benefits to the patient.

The occupational evidence comes mainly from medical radiologists, uranium and other miners working in inadequately ventilated mines in which there were high concentrations of radon, and radium luminisers, the people who used to paint luminous figures on watch and clock dials. Most of this evidence comes from people exposed many decades ago; improvements in radiation protection make it unlikely that such exposures will recur. There is limited evidence of radiation risk from workers in the nuclear industry. Despite extensive studies of this group and the relatively large numbers exposed, the numbers are still too low to enable statistically significant risk estimates to be derived. The results that are available, however, are not inconsistent with the risk estimates derived from other studies.

There have been many attempts to find correlations between cancer incidence and background radiation levels. The only statistically significant finding is an increase in lung cancer incidence among people living in houses with high radon concentrations, but only for people who also smoke. A more common finding is that areas with high levels of background radiation have lower numbers of cancer deaths that areas with lower levels. In the United States, high altitude states like Colorado have a significantly lower cancer death rates than the eastern seaboard states where radiation levels are lower by a factor of two. In five cities in India, cancer death and cancer incidence rates were found to decrease with increasing background levels. In China, detailed studies of large populations in two areas with a factor of two difference in background levels showed that cancer mortality rates were somewhat lower in the higher background area. These results do not mean that radiation is beneficial at these exposure levels, although this possibility cannot be ruled out, as discussed below, but strongly suggest that background levels of radiation are not a major cause of cancer.

Evidence for radiation-induced inherited abnormalities

There is good evidence from animal studies that radiation can cause inherited defects, but no similar evidence in humans has appeared at any dose level. Even in the Hiroshima and Nagasaki studies, no hereditary defects that can be ascribed to the radiation from the bombs have been observed in any of the children subsequently conceived by exposed parents, or among these childrens' children.

Evidence for beneficial effects

The evidence for a causal link between radiation exposure and increased cancer risk in humans is well-known, compelling and universally accepted. There is a less well-known but large body of evidence, from studies of a wide range of animal and plant species, that small doses of radiation can result in an increase in life expectancy. The mechanism that has been proposed to explain this is that small doses can cause changes in cells and organisms that make them better able to resist subsequent attack by harmful agents, including further doses of radiation, leading to improvements in overall health and expectation of life. The effect seems to operate by increasing the cell's capacity to repair damaged DNA.

The evidence for such adaptive responses to radiation was reviewed by UNSCEAR in its 1994 report. It found that: "Extensive evidence from animal experiments and limited human data provide no evidence to support the view that the adaptive response in cells decreases the incidence of late effects such as cancer induction in humans after low doses... As to the biological plausibility of a radiation-induced adaptive response, it is recognised that the effectiveness of DNA repair in mammalian cells is not absolute. The mechanisms of adaptation are likely to coexist with the mechanisms induced by low doses that may result in malignant transformations."

UNSCEAR concludes that: "An important question... is to judge the balance between stimulated cellular repair and residual damage. The Committee hopes that more data will become available and stresses that at this stage it would be premature to draw conclusions for radiological protection purposes."

It thus remains possible that radiation at low doses may be both harmful and beneficial, but in different ways.

Risk estimates for stochastic effects

The measures most commonly used to quantify the harm associated with the stochastic effects of a radiation exposure are:

- for cancers, the probability of dying prematurely of cancer as a result of the exposure;

- for hereditary effects, the probability of a severe inherited abnormality occurring as a result of the exposure. .

It may be possible to express the risk per unit dose, so that the effects of different dose levels can be compared on an equal footing, but this assumes a linear relationship between the size of the dose and the probability of the consequence occurring. There is reasonable evidence for such a relationship over a limited range of doses, mainly from the Hiroshima and Nagasaki studies. A linear relationship is also consistent with the results of animal and cellular studies, again over a limited dose range, provided the radiation is in the form of beta, gamma or X-rays.

The question of the shape of the dose-response curve, which describes the relationship between the dose and the probability of harm, is a crucial one. Most of the data comes from doses of 1 Sv or more, or down to 50 mSv in a few cases, but, except for patients receiving radiotherapy, most people receive doses well below 1 Sv in total during their whole life. Furthermore, most of the evidence comes from doses delivered over a short period, unlike natural and most man-made exposures, again other than radiotherapy. Is the risk associated with a dose of 1 or 2 mSv, a typical annual dose from the natural background, one thousandth of that known from observation to be associated with a dose of 1 or 2 Sv? Does a dose of one millionth of a sievert, a typical annual dose to members of the public from the nuclear industry, carry any risk at all? The concept of collective dose, commonly used to describe the overall consequences of activities involving radiation exposure, depends on linearity over the whole range of doses experienced as a result of the activity, including very small doses spread among very large numbers of people. To what extent is it sensible to ascribe any risk to doses that are a small fraction of those received from the natural background, when large variations in this background do not appear to be associated with any harmful effects?

These questions are unlikely ever to be answered by epidemiological studies, since the size of the study population needed to give statistical confidence is impracticably large. The only guidance that is currently available comes from the observed shape of the dose-response curve at relatively high doses and dose rates and from knowledge of the way in which radiation is able to cause cancer and inherited abnormalities, coming from theoretical and laboratory studies. It is now generally agreed that for the purposes of recommending and implementing measures to protect people from radiation, when it is clearly right to err on the side of caution (the so-called precautionary principle), the risk estimates derived from such studies should be used, together with the assumption that any dose, however low, carries some risk. There is also general agreement that risk estimates derived from studies at high doses and dose rates should be reduced by a factor of around two when being used to derive risk estimates for low doses and dose rates. For this purpose, a low dose is defined as one below 0.2 Sv and a low dose rate, for higher doses, as one that is received as less than 0.1 Sv per hour.

ICRP's latest risk estimated for stochastic effects are summarised in Table 1. The table also gives estimates, based on these figures and the assumption of a linear dose-response relationship, of the risks of fatal cancer and serious hereditary damage associated with natural background radiation and doses to the public from nuclear power production. In the case of hereditary effects, the negative findings at Hiroshima and Nagasaki suggest that the figures are unlikely to underestimate the actual risk.

Table 1: Excess risk of stochastic effects

Excess risk of stochastic effects for the general population resulting from various radiation doses, assuming exposure at low doses and dose rates.

DOSE	EXCESS RISK	
	Fatal cancer[††]	**Severe hereditary effects**
	%	%
1 Sv	5	1.3
2.4 mSv*	0.012	0.003
0.1 mSv**	0.0005	0.00013
0.001 mSv[†]	0.000005	0.0000013

* The average annual dose from natural background radiation

** A typical maximum annual dose to people living near nuclear installations

† The average annual dose from nuclear power production

†† If the "normal" incidence of fatal cancer were 25 per cent, a typical figure for industrialised countries, a dose of 1 Sv would increase this to 30 per cent and a dose of 0.001 mSv would increase this to 25.000005 per cent

Chapter 5

RADIATION PROTECTION

International recommendations aimed at protecting people from the harmful effects of radiation have existed since the 1920s. Since that time recommendations have been developed by several bodies, particularly the International Commission on Radiological Protection and the International Atomic Energy Agency.

Protection recommendations are based on three main principles. First that any activity resulting in radiation exposure should be justifiable in terms of the benefits of that activity. Second that measures should be taken to keep exposure at a level as low as is reasonably achievable. Third that specific limits should be set for the maximum dose to which any individual will be exposed as a result of the activity.

The current recommendations define a "practice" as any activity that may result in increased exposure to an individual. Limits are set on the exposure that should result from any given practice, both for workers and for the general public. Medical practices are not subject to such limits, but exposures should be as low as practicable, consistent with medical need.

Recommendations are made for actions that should be taken in order to avert or reduce exposure below certain levels following an accident, or in certain chronic exposure situations which may be natural (for example high radon levels in dwellings) or man-made (for example when cleaning up areas contaminated by past activities). Such actions are called "interventions". Levels are set at which different action should be considered, for example evacuation should be considered in order to avert individual doses of between 50-500 mSv. These levels are known as "reference levels for interventions". The recommendations acknowledge that there is a range of other factors that need to be taken into account before the action most appropriate to a given circumstance can be determined.

In practice, protection of the public is ensured mainly through processes such as site licensing and discharge authorisation. The necessary level of protection is achieved by containment of radioactivity, using multiple barriers and multiple safety systems, and treatment of any radioactive effluents. The protection of workers is also based on containment, and in addition involves optimising four main parameters: the radioactivity of the source, the duration of any exposure, distance from the source, and shielding. Workers should be suitably trained and qualified, and aware of the risks likely to be encountered.

The effectiveness of radiation protection measures needs to be confirmed by monitoring exposures to individuals and measuring the general levels of radiation and concentrations of radioactive materials in the environment. It is usual practice to monitor directly the workplace exposure of individuals in relevant occupations, for example medical radiographers or workers in the nuclear industry. Direct monitoring of members of the public is less common, instead exposures are estimated using models based on environmental measurements and knowledge of the various pathways by which radioactive materials entering the environment can reach people.

The development of protection measures

The need to protect people from harmful effects, both early effects such as skin burns and late effects such as cancers, was recognised soon after the early discoveries of radiation. Achieving such protection was hampered, however, by the difficulty of measuring the radiation doses that people were receiving. Instruments to measure dose were not developed until the 1920s. Before that, various devices were used, such as crystals which changed colour or fluoresced on being irradiated, or photographic plates. But many different types of X-ray tubes were being used and operating conditions varied from machine to machine and even from day to day, so cross-calibration and standardisation were almost impossible. For want of a better measure, the rather alarming concept of "erythema dose", the dose required to produce a visible reddening of the skin (erythema), was introduced.

The first formal step to control radiation exposure of workers appears to have been taken by the Deutsche Röntgen Gesellschaft, which issued a leaflet in 1913 making specific recommendations on the thicknesses of lead screening to be used when working with X-rays. In 1915 the Röntgen Society in Britain agreed "that in view of the recent large increase in the number of X-ray installations, this Society considers it a matter of the greatest importance that the personal safety of the operators should be secured by the universal adoption of stringent rules."

Experience with X-rays during the 1914-1918 war, often in the difficult conditions of field hospitals, led to a further realisation of the need for protection. Several countries introduced protection measures in the succeeding years. In 1925, the International Congress of Radiology, meeting for the first time in London, agreed on the need to develop a unit for measuring X-ray exposure and on the importance of establishing internationally agreed standards for working with radiation. This Congress set up the International Commission on Radiation and Measurements (ICRU) to review and advise on the problems of radiation measurement; the body still fulfils that role. At its second meeting in Stockholm in 1928, the Congress issued a number of recommendations on protection and agreed to form an International X-ray and Radium Protection Committee, later renamed the International Commission on Radiological Protection (ICRP), to keep these recommendations under review. ICRP's recommendations for the protection of workers and the public continue to form the basis for radiation protection standards world-wide. Early protection recommendations were based on the concept of there being a threshold level of exposure below which no damage occurred. In 1925, scientists in Germany and Sweden suggested, independently, that an "acceptable exposure" for radiation workers was about one tenth of an erythema dose per year. In 1934, ICRP recommended that a person in normal health should be able to "tolerate" a dose of around 0.7 Sv per year, roughly equivalent to the one tenth of an erythema dose suggested earlier. Even in those early days, however, ICRP warned that the worker "should on no account expose himself unnecessarily" and "should place himself as remote as practical from the X-ray tube", ideas very close to today's recommendations.

The discovery by Müller in 1927 that X-rays could accelerate mutation rates in fruit flies led to a realisation that radiation could cause genetic defects as well as cancers. Further experiments with fruit flies showed that mutation rates remained linearly proportional to dose as the dose was reduced, leading to doubts about the existence of a threshold for stochastic effects. In the early 1940s, the Advisory Committee on X-ray and Radium Protection in the United States discussed the possibility of reducing the recommended tolerance dose level by a factor of ten because of the risk of genetic damage.

Further research, in particular the large amount of information from the nuclear weapons programme during the Second World War and studies of the devastating effects of the Hiroshima and Nagasaki bombs, resulted in ICRP making major changes in its recommendations after it was re-formed in 1950. The maximum permissible dose was reduced to about one fifth of the pre-war level. "Tolerable dose" was replaced by "permissible dose" to reflect doubts about the existence of a threshold. ICRP recommended that "every effort be made to reduce exposure to all types of ionising radiation to the lowest possible level".

This caution was reflected in ICRP's 1958 recommendations, which were based on the assumption that there was no threshold for stochastic effects, although it stressed that this assumption was not supported by any evidence. All ICRP's subsequent recommendations have been based on the same fundamental assumption, which implies that any level of radiation exposure, however low, carries some risk. Once absolute safety cannot be guaranteed, even at very low levels of exposure, it becomes necessary to make judgements about the acceptability of the risks that are assumed to exist. The problem is not a new one, since there are few human activities that are absolutely risk-free. What constitutes an "acceptable risk" is very difficult to define; there are professors, entire university departments and even special institutes grappling with the problems of risk assessment, perception, comparison, communication and acceptance, producing a steady flow of research papers and books, and there are frequent national and international conferences on these subjects. The public often appears to want risks to be reduced to zero, except when they have to pay for such reduction themselves or when the risks are from voluntary activities such as smoking and driving. The problems are addressed further in Chapter 6.

International recommendations

The current aim of national and international bodies concerned with radiation protection is to prevent harmful deterministic effects (*i.e.* those associated with cell killing and exhibiting a threshold) and to limit the probability of stochastic effects (*i.e.* cancers and genetic damage, assumed to be linearly related to dose, without a threshold) to levels deemed to be acceptable. The approach to the problem of acceptability was for many years based on comparisons with other risks to which workers and the general public are exposed. These cover a very wide range, from the very low risk of being struck by lightning to the very high risks accepted voluntarily by some people, for example those associated with some sporting activities, such as rock climbing.

Current recommendations are based on absolute rather than comparative judgements of what constitutes an acceptable risk. At the upper end of the risk scale, there is broad agreement that exposing anyone to an activity that carries an annual risk of death of over one in a thousand is unacceptable, however beneficial the activity is. An exception is some medical treatments which may carry higher risks, but these are justified by the even higher risk of not giving the treatment. Also, higher risks are sometimes accepted voluntarily. An annual risk of one in ten thousand is not uncommon in industry; it is also approximately the average annual risk of dying in a traffic accident in a typical OECD country. An annual risk of death of around one in a hundred thousand appears to be tolerated by most people, provided the activity that gives rise to the risk is associated with some benefit. Few people have any concerns about annual risks that are below one in a million, although some activities that carry risks of that order may still be under regulatory control, and there may be a requirement to reduce such risks further if possible.

The situation is more complex when considering accidents that involve large numbers of people. For example, there is generally far more concern about a single accident, say a plane crash, that kills 100 people than about the continuing toll of road accidents, which cause far more deaths per year than air traffic.

Against this background, the recommendations of ICRP and other bodies have, since 1977, been based on three main principles:

Justification. A practice involving radiation exposure should do more good than harm – the overall benefit from the activity should offset the possible adverse effects from radiation that it causes. In some cases the principle is easy to apply, for example it has led to the banning of toys and jewellery containing radioactive materials. In others, and certainly in the case of nuclear power, balancing benefit against harm is more difficult and may require consideration of complex economic, social, environmental and political issues.

Optimisation of protection. For any source of radiation exposure, the doses, the number of people exposed and the likelihood of being exposed should be kept as low as reasonably achievable, often therefore called the "ALARA" principle. The word "reasonable" is included because a point must come where the cost of achieving a further reduction of an already low risk cannot be justified. For example, all motorways have central crash barriers but no-one suggests having them down the middle of quiet country roads, even though some lives may be saved as a result. Economic and social factors must be taken into account when assessing what is reasonable.

Individual dose and risk limits. The need to ensure that no individual is exposed to an unacceptable level of risk requires the setting of dose limits. The limits do not apply to patients irradiated for medical reasons, since their exposure is decided on the basis of therapeutic or diagnostic indications which will provide a direct benefit to the exposed individual. Exceeding a dose limit does not automatically result in serious harm, any more than exceeding a speed limit automatically leads to a serious road accident – the recommended dose limits are far below the threshold for deterministic effects.

Practices and interventions

ICRP's 1990 recommendations, while based on the general principles first set out in 1977, reflect more closely the networks of events and situations that result in human exposure. Each part of a network starts from a source of radiation. The environmental pathways that lead to the exposure of one or many individuals cover a wide range; they may be simple, as in the case of the inhalation of radioactive dusts by miners, or very complex, as in the case of uptake of fallout from nuclear testing, following deposition on the ground and subsequent contamination of animals and foodstuff. People are generally exposed to many sources of radiation, and protection measures can be applied both to the sources themselves and to the individuals who are receiving the doses.

PRACTICES AND INTERVENTIONS

PRACTICES are activities that increase radiation exposures, for example nuclear electricity generation or industrial and medical uses of radiation or radioactive materials.

INTERVENTIONS are activities that seek to reduce radiation exposures, for example reducing radon levels in homes or cleaning up accidental radioactive contamination.

Activities are now categorised as:

Practices. These are activities that increase the radiation exposures that people already receive from the natural background or existing sources, or increase the likelihood of people being exposed. When a practice is implemented, it may increase the radiation exposure of some individuals, because it introduces a new source or modifies a pathway, or it may increase the number of individuals exposed. Practices result in additional exposures that can be foreseen and protective measures can be applied when the activities are being planned. Examples of practices are the use of radiation in medicine for diagnosis and therapy, sterilisation of medical materials, industrial radiography, food irradiation, smoke detectors and the operation of the entire fuel cycle for nuclear electricity production.

Interventions. These are activities that seek to reduce a radiation exposure that is judged to be unacceptable, from a source that already exists and for which protection was not planned or could not be planned. An intervention may involve removing the radiation source, modifying a pathway or reducing the number of exposed individuals. Interventions can reduce exposure from a chronic situation, for example high radon levels at home or residual radioactivity from past activities or events, or from an emergency situation, for example following an accident.

All three protection principles apply to the introduction of new practices, and national regulations generally require that the appropriate authorities are satisfied that the practice is justified, protection is optimised, and dose limits will not be exceeded before granting any construction and operating licenses that may be needed.

The situation for interventions is different. Interventions are applied to people or their environment rather than to the source of radiation exposure, which already exists. The protective measures aim to reduce doses to acceptable levels, but may carry their own risks and disadvantages. The principles of justification and optimisation still apply, but in different ways. The intervention should be justified, in the sense that it should do more good than harm, with due regard to health, social and economic factors, and it should be designed and implemented in the most effective way possible. In other words, the benefits of the reductions in dose achieved by the intervention, less the total disbenefits, in terms of costs, increased risks and other disadvantages, should be as large as reasonably achievable. For interventions, however, dose limits are irrelevant: the source of exposure already exists and a decision is needed on whether an intervention is justified. To guide those who have to decide whether to implement interventions, ICRP has set general principles and IAEA has recommended a series of reference levels. These relate to the circumstances of the event or the chronic situation for which intervention is required and the type of intervention being considered, for example sheltering, evacuation or banning of certain foodstuffs after an emergency, or taking measures to reduce high radon concentrations in buildings.

Dose limits for practices

Workers

Dose limits for practices are based on the best available data on the effects of radiation and judgement of risk acceptability. For workers, the recommended dose limit is 20 mSv a year, averaged over five consecutive years of exposure, with no more than 50 mSv being received in any single year. The risk to a worker regularly exposed at the limit throughout a 40 year working life is deemed to verge on unacceptability. It corresponds to a total lifetime risk of fatal cancer of about 5 per cent, over and above the "natural" risk of about 25 per cent. This level of risk lies at the upper end of the range of occupational risks experienced in most developed countries.

The application of the ALARA principle, together with the fact that radiation workers are seldom exposed to high dose levels throughout their working lives, results in the actual radiation exposures of workers being generally well below the 20 mSv a year limit. For example, the average worker dose at nuclear power stations is around 2.5 mSv a year, close to the average annual natural background dose, and the corresponding total lifetime risk of fatal cancer, assuming exposure at the average dose for a whole working life, is about 0.6 per cent. Average doses to medical radiation workers are around 0.5 mSv a year. Aircrew get typically between 2 and 3 mSv a year from the enhanced levels of cosmic radiation at high altitudes. Mine workers average 1 to 2 mSv a year in coal mines, 4 to 5 mSv a year in uranium mines and 1 to 10 mSv a year in other mines. Table 2 shows annual world-wide collective doses for a range of occupations. There are wide variations in occupational exposure levels from country to country, depending mainly on the type of nuclear installation (reactor type, underground or open-cast uranium mine) and on the stringency of the national regulations.

Public

For the public, the recommended dose limit for exposures resulting from current practices is 1 mSv a year or, in special circumstances, 5 mSv in a single year provided that the average dose over five consecutive years does not exceed 1 mSv a year. This limit is more restrictive than that for workers for a number of reasons. It applies to a much larger number of people and throughout life rather than only for a working lifetime, and the public at large includes children who are somewhat more susceptible to radiation than adults. An exposure of 1 mSv per year throughout a 70 year life increases the risk of fatal cancer from the "natural" level of about 25 per cent to about 25.4 per cent.

Table 2: All occupations: annual collective doses

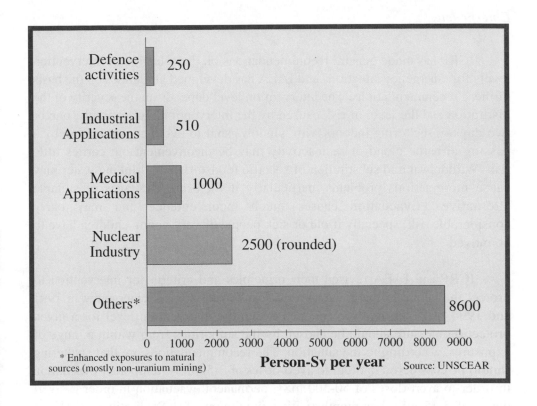

The way that the public dose limit is applied, and the ALARA principle, results in doses that are usually well below, and for the vast majority of people, very far indeed below the limit. The limit is applied to the group of people receiving the highest doses from the practice, called the critical group. For example, for discharges to the Irish Sea from the Sellafield reprocessing plant in Britain, the critical group is a small number of people living near the plant, probably fewer than ten in total, who eat large quantities of locally caught seafood. These people are estimated to receive about 0.15 mSv a year from this source. The average dose to the entire United Kingdom population from all discharges from all nuclear installations, of which Sellafield makes the largest contribution, is less than 0.0003 mSv a year, giving a lifetime risk of fatal cancer of around 0.0001 per cent. Similar patterns of dose are found at other nuclear installations and in other countries, with small doses to local populations and negligibly small doses further away.

Reference levels for interventions

Emergencies

ICRP has made general recommendations on the selection of intervention levels for emergency situations and IAEA has developed guidelines on the basis of these recommendations. The intervention level depends on the severity of the disruption and the level of risk caused by the intervention. For example, one or two days of sheltering indoors with windows shut to avoid contamination by a passing airborne cloud of radioactivity may be inconvenient but carries little risk. Withdrawal and substitution of specific foodstuffs and drinking water may cause more serious problems, particularly if there are shortages of suitable alternatives. Evacuation causes much inconvenience and may carry considerable risk, specially if old or sick people or very young children have to be moved.

ICRP and IAEA revised their principles and criteria for intervention to protect the public in the event of a nuclear or radiological emergency in 1991 and 1994. ICRP currently recommends that the intervention level for a given protection measure should be chosen by the authorities from within a range of exposures, according to the situation. The recommendations are that sheltering should be considered in order to avert doses of 5-50 mSv, temporary evacuation in order to avert doses of 50-500 mSv, permanent evacuation in order to avert doses of 5-15 mSv per month (with a maximum of 1 Sv lifetime dose) and iodine prophylaxis (taking non-radioactive iodine tablets to saturate the thyroid gland in order to prevent absorption of radioactive iodine) in order to avert thyroid doses of 50-500 mSv.

IAEA has defined levels of dose, applicable for most situations, at which interventions should be implemented; they are given in Table 3. In some specific situations the levels may be modified, depending on the particular factors prevailing when the emergency occurs. The intervention should be implemented when the dose which may be averted by the protective measure is higher than the intervention level recommended for that measure. The IAEA levels are intended to be starting points in the process of deciding on whether to apply intervention measures following an emergency; the final decision would have to take account of factors such as the age or health of those affected, transport limitations, weather conditions and compounding hazards (*e.g.* hazardous chemicals).

70

Table 3: Intervention levels

INTERVENTION MEASURE	INTERVENTION LEVEL
ACUTE EXPOSURE: *all circumstances*	
Urgent protective action to avoid serious injury	1 Gy (whole body) in less than 2 days
CHRONIC EXPOSURE: *emergencies*	
Sheltering	10 mSv in no more than 2 days
Temporary evacuation	50 mSv in no more than 1 week
Temporary relocation (initiation)	30 mSv in 1 month
Temporary relocation (termination)	10 mSv in 1 month
Permanent resettlement	1 Sv in a lifetime
Iodine prophylaxis	100 mGy absorbed dose to thyroid
CHRONIC EXPOSURE: *non-emergency situations*	
Radon in dwellings	200-600 Bq m^{-3} of radon-222 in air (equivalent to 3-10 mSv per year assuming 7 000 hours per year indoors)
Radon in workplaces	1 000 Bq m^{-3} of radon-222 in air (equivalent to 6 mSv per year assuming 2 000 hours per year at work)

Source: IAEA

Chronic exposure situations

A complex set of considerations is involved in deciding on whether intervention is justified in the case of chronic exposure situations. These include the individual and collective dose levels being experienced, and the risks, benefits, financial and social costs, and financial liability for the remedial actions.

Currently, numerical guidance is only available for chronic exposure to radon. For example, ICRP and IAEA have recommended that action should be taken to reduce radon levels in homes when its concentration is such as to result in doses in the range 3-10 mSv per year. The ICRP recommendations emphasise the role of national authorities in deciding on the levels of funding for general reductions in radon levels or other aspects of housing improvements.

There is an interesting contrast between what happens in practice in emergency and chronic exposure situations, particularly when the source of chronic exposure is natural. Experience in high radon areas in many parts of the world suggests that there is a low level of interest in the problem, even when brought to the attention of the public, even though lifetime doses from radon in such areas can be well over 1 Sv. This is the level at which, according to the guidelines for intervention following an emergency, permanent relocation should be implemented. If the emergency intervention standard were to be applied to high radon areas, many of them would now be depopulated. This suggests that public concerns are governed more by the nature of the source of the radiation than by the magnitude of the exposures received.

Protection in practice

The protection of the public is ensured essentially through processes such as licensing and discharge authorisation, which are designed to ensure compliance with protection principles. The necessary protection is achieved by containing the vast majority of the radioactivity, using multiple barriers and multiple safety systems if required, throughout the life of facilities and during any subsequent storage, transport and disposal operations that may be necessary. Any radioactive effluents are treated so that only small amounts of radioactivity are released into the environment. People who live near large installations such as nuclear power stations are protected from direct radiation mainly by the massive structures used to contain the radioactive cores and ancillary buildings. As shown in Chapter 3, such measures have, with a few notable exceptions, resulted in doses that are well within the prescribed dose limits and a small fraction, usually a very small fraction, of the doses received from natural sources.

Medical diagnostic exposures vary considerably from country to country, reflecting, in part, different levels of health care, and even from hospital to hospital within one country. There is a need to initiate or pursue the implementation of an ALARA approach to diagnostic radiology, which is responsible for a significant proportion of overall collective doses due to medical practices.

The protection of workers directly exposed during their work is subject to the same overall protection principles of justification and optimisation, taking into account the cost of protection and all other relevant parameters. Generally, higher dose limits are applied. An important requirement is that the workers are suitably trained and qualified, and that they aware of, understand and accept the risks likely to be encountered.

Reduction of exposure is based on the choice and design of equipment adapted to the working conditions in a radiation environment and the use of procedures which aim to reduce individual and collective doses. The reduction of individual doses is achieved through optimisation of four main parameters:

- the radioactivity of the source
- the duration of the exposure
- distance from the source
- shielding from the source.

Source radioactivity

The first stage in implementing protection measures is to identify the nature and type of radiation emitted by the source. Radiation from a source outside the body (external radiation) can be sufficiently penetrating to reach the individual, for example gamma radiation, X-rays or neutrons. Internal exposure from radioactive materials which may contaminate the environment, for example alpha emitters, can best be avoided by appropriate means of containment, such as glove boxes; generally collective protection is sufficient and does not need to be supplemented by individual specific protection. In all cases, good practice requires that minimal radioactivity should be used; where a choice is possible between different sources, the one resulting in the least exposure should be used.

Duration of exposure

The less time is spent in the vicinity of a radiation source, the less radiation dose will be received. This time should therefore be reduced to the lowest level, taking into account all the different phases of the overall operation. Since the total exposure is the integration of the dose rate, which may vary over the time of the exposure, operators must be aware of the source characteristics, including any directionality of the radiation field and the dose rates at various distances. Appropriate job preparation and work organisation are important ways to minimise the duration of exposures. However, excessive speed can cause mistakes, resulting in greater exposure. Training and practice, using simulators in a non-radiation environment, are often used to minimise the time taken for specific operations.

Distance

As radiation travels away from its source it spreads out and becomes less intense; increasing the distance from a source therefore decreases the amount of radiation received by exposed individuals. For a small source the exposure decreases as the inverse square of the distance from the source. For example for a given dose rate at one metre, the expected dose rate will be divided by four at two metres, by one hundred at ten metres and by four hundred at twenty metres. Distance is commonly used to reduce exposures to acceptable levels throughout industry and medicine. For example, the use of long control cables, the control of access around radiation sources and manipulation of small sources with long pliers all result in considerable reductions in doses to the hands and the whole body.

Shielding

Shielding from radiation can be provided by a sufficient thickness of material – from a thin piece of paper or plastic for alpha radiation to a thick wall of concrete for neutrons and gamma rays. In general, the denser the material, the more effective the shielding. The radiation is absorbed by the shielding material but it can also be scattered in various directions and shielding design has to take such scattered radiation into account.

Since the efficiency of the shielding depends on the density of the material used, uranium, the densest natural material, is the most effective, except when the radiation is in the form of neutrons which can interact with the uranium and produce further radiation. Other metals such as lead, tungsten and steel are commonly used. Concrete is not as effective as these metals, but it is often used because it is comparatively inexpensive and easy to use in construction. A thick wall of concrete is as effective as a much thinner wall of lead.

Shielding can take many forms. Medical and industrial radiography makes much use of collimators: small pieces of lead or other metal surrounding the source to absorb radiation not directed at the individual or object being radiographed. When radiation is used at a permanent facility, such as a hospital or an industrial installation, thick walls of concrete are built around the irradiation room for shielding, with a labyrinthine entrance if necessary to absorb scattered radiation. In the nuclear industry, reactors are surrounded by concrete several metres thick; used radioactive fuel is immersed in pools of water, five metres or more deep, which shields the personnel from the intense gamma radiation emitted by the fission products in the fuel.

Protection often requires combinations of these various techniques. In addition, when there is a possibility of spread of contamination, containment is provided to ensure air- and water-tightness between the source of radiation and the working areas; frequently the structures provide both shielding and containment. In nuclear reactors, there are strict barriers between radioactive fuel, coolant fluid circuits, working areas and the external atmosphere. Air in working areas is renewed by ventilation systems which filter it continuously and capture radioactive particles on the filters. Some operation may require special protective clothing and other equipment to prevent contamination from reaching the body surface and the respiratory tract and to avoid its spread through and outside the working area or the facility. The wearing of protective clothing must not pose unnecessary risk from conventional safety hazards.

The costs of the various measures used to protect workers and the public can cover a very wide range, from a few dollars for simple screens to protect radiographers in medicine and industry to hundreds of millions of dollars for some effluent clean-up plants. The cost of protection is discussed in the next chapter.

Monitoring

The effectiveness of measures to protect people from radiation needs to be confirmed by monitoring radiation exposures and radioactivity in the environment. Although radiation cannot be felt, seen, heard or smelt, it is easily detected by relatively simple instruments and can be measured with high accuracy at very low levels.

Radiation monitoring is extensively used wherever radiation is likely to be a threat to human health. Personal monitors are used to determine the doses received by individual workers from external and internal sources of radiation. Environmental monitors are used to measure radiation levels in workplaces or near installations using radiation or radioactive materials. These may be connected to alarm systems which operate when a certain level of radiation is exceeded and may trigger automatic safety systems. In some countries, information about radiation levels around nuclear installations is continually available, for example in France on the public Minitel system. In others, such information is regularly published, and in particular made available to local communities or their representatives.

Several types of detector are available to measure radiation doses from external sources. They are worn by workers likely to receive doses that are significantly higher than those received by members of the public and are used to provide complete records of the magnitude and types of radiation exposure.

Radiation doses can also result from radioactive materials that are inhaled or ingested, or absorbed through the skin. The uptake and retention of radioactive materials depends on many factors such as their physical and chemical properties. The amount of radioactivity in the body can be estimated by direct measurement in the body and by measurement of biological samples. The total dose is then assessed using bio-mathematical models to follow the movement of radioactivity within the body and calculate doses to various organs.

For workers, monitoring provides the necessary knowledge for maintaining a safe and healthy environment in the workplace and its surrounding area. The principal objectives are to support the control of operations and facilities, to demonstrate compliance with dose limitation requirements and to detect unexpected occurrences. Monitoring of workers also provides data to guide the medical response to actual or suspected contamination incidents or exposures requiring such response. It also serves other functions unrelated to the control of exposure, such as the provision of dose data for epidemiological studies. Dose records are kept throughout the professional lives of workers.

In general there is no need for regular monitoring of public exposures. The dose levels from man-made sources are usually immeasurably small, especially when compared with doses from natural sources. They are usually estimated using models based on knowledge of the quantities of radioactive material entering the environment and the pathways from the source to people (see page 38). In some cases where abnormally high levels of radiation are suspected, for example following a serious accident, or for chronic situations such as high levels of radon in the home, monitoring can be used to assess doses and provide information on the need for intervention.

Chapter 6

SOCIAL AND ECONOMIC ISSUES

Radiation protection measures must be based on a scientific understanding of the risks of radiation exposure and an economic analysis of the costs involved in reducing such risks. It is not, however, sufficient to consider such factors alone. It is also necessary to address fundamental issues relating to society's attitude to risk in general: such issues as equity, fairness, and compensation.

Public perception of risk is at least as important as technical risk assessment. Studies into public attitudes to risk consistently demonstrate a level of public concern about some man-made sources of radiation, particularly nuclear power, that is far greater that would be expected from an analysis of technical factors alone.

Risk reduction costs money. Society needs to make decisions on how resources should be allocated to the management of risk. In order to decide between different ways of allocating limited resources, some kind of measure of benefit is required. Generally, this involves relating cost to benefit in terms of lives saved. In practice individuals and society are constantly balancing cost and benefit in this way, for example when considering investment in health or transport.

Ultimately decisions relating to radiation protection must be made within a framework of public acceptability and accountability. Communication and consultation are essential to this process.

———————◆▶———————

In the early days of radiation protection, regulation of exposure was simple. Activities were permitted if they resulted in doses below a certain level, at which they were assumed to be harmless, otherwise they were forbidden. Today's principles of justification and optimisation are much more difficult to apply. On what basis can one compare benefits with harm? How can one compare radiation-related benefits and harm with those of other practices? How can one ensure a fair balance of risk between individuals and larger numbers of people, and between workers and the public? How much is it reasonable to spend on protection measures? Who should take such difficult decisions? To what extent should public opinion be taken into account and to what extent should the public be involved in the decision making process? The ethical, social and economic dimensions of these questions are addressed in this chapter.

Ethical issues

Evaluations of radiation risk and the development of measures to protect people are based mainly on technical judgement and assessment. Such an approach is a necessary first stage in decision making, but it is not sufficient. Additional analysis is needed, for example relating to the uncertainties of risk assessment, the acceptability and tolerability of risk, its controllability, equity and fairness regarding the distribution of risks and benefits, and the question of compensation for exposure to risk. These analyses belong to the world of ethics.

The use of radiation has provoked public controversy and many beliefs which are conceptually unrealistic or contrary to observation are put forward on ethical grounds. Examples are "zero risk" targets, an unwillingness to ascribe a finite value to human life for decision-making purposes, and an absolute intolerance of risk whatever its level and associated benefits. These questions are not specific to radiation, but are probably exacerbated by the particular perception of radiation risk by the public. In addition, radiation has to some extent become a paradigm for all other potentially harmful agents, and the advanced state of knowledge of its detrimental effects and the highly developed safety and protection frameworks themselves probably contribute to the level of public concern.

As with many other human activities, the risk of radiation is not uniformly distributed. Those exposed to a higher risk generally do not receive a higher benefit; for example a member of a critical group near a nuclear power station will not get more advantage from the electricity produced than any other member of the public. Exceptions are patients who receive direct benefit from diagnostic or therapeutic exposures. This general situation is currently well accepted, probably because it is similar to many other human activities, especially when dealing with occupationally exposed workers compared with the general population.

The question becomes more difficult when the inequity in distribution involves future generations. The possible transmission of damage to progeny is often of particular concern, although this problem is not specific to radiation;

many chemicals can have similar effects. On the other hand, there is little appreciation that radiation is probably the first toxic agent for which plans and provisions are being made for the long-term future. Arguments about the uniqueness of the long-term hazard from radioactive materials are false: although some are very long-lived, their radiotoxicity eventually falls to zero, whereas some non-radioactive pollutants remain toxic for ever.

Some of the problems of distribution of risk and benefit raise fundamental questions about the benefits of technical progress, trust in the ability of future generations and the evolution of the environment and social and economic structures. Nevertheless, since one cannot wait for all uncertainties about the long-term future to be resolved, such difficulties do not remove the need for action. Decisions can only be taken within the general framework of an implicit social contract of sustainability, aimed at providing future populations with similar opportunities to those available to today's.

Many decisions involve individual consent to societal risk. Attitudes to societal risk vary widely. At one extreme, prohibition of a beneficial activity that may carry some societal risk is unrealistic and often detrimental overall, given the unattainability of zero risk. At the other extreme, one cannot assume that society will accept a risk just because it has accepted some other equally or more risky activity. For example, imposed risks, such as those of nuclear power, and voluntarily accepted risks, such as driving a car or smoking, are associated with different levels of distributive equity, benefits, profits, etc. The triggering parameters for the NIMBY syndrome ("not in my back yard") are linked more with equity than with the potential risk, uncertainty of the future, or fear. Provided that the relevant factors are properly considered, however, comparisons between risks of different origins can provide some guidance, for example comparing societally imposed radiation risks with natural background levels. In the final analysis, decisions about issues such as absolute safety and voluntariness are matters of people's democratic rights to self determination. The scientific inputs to decision-making are important but not dominant, since questions of societal values are involved.

Another contentious area is compensation for risks. The general practice in developed societies is to offer compensation only for the alleviation of a specific detriment, such as the development of a disease or loss of property or work. Normally, there can be no compensation offered for an increased risk of cancer before it has actually developed. In other words, if a level of imposed radiation exposure, occupational or public, is judged to be unacceptable, then the moral way would be to reduce the exposure to an acceptable level, and certainly not to offer compensation. In fact the levels of risk encountered in various human activities are rarely directly linked with any form of compensation. Experience in the Ukraine, Belarus and Russia after the Chernobyl accident, where compensation has been widely distributed to affected populations, prompts the question of whether individual benefits (from compensation) outweigh societal harm, particularly when limited resources are available.

Many of the above questions are considered, more or less implicitly, in the current general framework of radiation protection, with its emphasis on justification and optimisation of protection, both for practices and for interventions.

Some other ethical issues, which are often, but wrongly, considered as specific to radiation, are more difficult. One important issue is the possibility that some people may have an enhanced susceptibility to radiation – indeed the enhanced susceptibility of the foetus to radiation is already taken into account in the case of potentially pregnant women radiation workers. In fact, the problem is a very general one, since humans are not uniformly susceptible to toxic agents. Increasing understanding of the human genome and its repair capabilities have provided basic knowledge which may allow risks to individuals with different radiation sensitivities to be differentiated. This might affect matters such as employment policies and intervention strategies in the case of accidents and may raise fundamental question of selection and discrimination which are in contradiction with currently accepted ethics. As research in this field is at an early stage, the question is not yet directly answerable. However, it might be one of the most important issues for the future and might transform the generic approach to fitness to work.

Risk perception

The processes by which decisions relating to potentially risky activities are taken by democratic societies are complex. They are often influenced by the public's perception of risk, in particular when this differs substantially from technical assessment by scientists and engineers. The formal decision process for activities involving radiation exposure generally follows a set pattern:

- International bodies such as ICRP and, in some cases, national bodies, estimate the risks on the basis of the best available information and make recommendations for protection measures; such bodies are not generally concerned with making judgements about the relative importance of different kinds of risk to society or with the management of risks.

- National bodies review and, usually, endorse the recommendations.

- Regulatory agencies, usually national but sometimes, as in the European Union, also international, issue specific, legally enforceable requirements based on the recommendations.

The review and regulatory phases of this process may be strongly influenced by public perceptions of risk, which often differ markedly from the measured or estimated risks.

Many factors influence public perceptions. Familiar, well understood, voluntary, reversible, "fair" risks, and risks associated with clearly beneficial activities are often seen as acceptable, while those that affect identified individuals, children or future generations, and whose benefits are unclear or inequitably distributed, are not. For example, there is little public pressure to reduce radiological risks in hospitals although dose reductions could be achieved at relatively low cost – the activity that gives rise to the risk is universally seen as beneficial. Recent media coverage, the degree of trust in the institutions seen as responsible for the risk and its regulation, the availability of apparently safer alternatives, the possibility of misuse (sabotage, terrorism), all influence perceptions. An important factor is whether the risk is natural or man-made. Radon gives rise to little public concern despite the large numbers of people, including children, that are exposed; the public perception of this risk is so relaxed that radiation risks that are ten times higher than would be tolerated from nuclear installations (10 mSv rather than 1 mSv a year) are accepted as reasonable. Accidents which may affect large numbers of people are seen as unacceptable however low the probability of occurrence, while accidents which occur very frequently but only affect one person or a few people at a time (*e.g.* road accidents) are virtually ignored, by all except the unfortunate victims.

The risk of an activity is seldom considered in isolation. People judge activities as a whole. Nuclear power, in particular, is seen as impersonal, highly centralised and unfriendly, run by a technological elite, a paradigm for what many people find unacceptable in advanced industrial societies. The risks are associated with those of nuclear weapons, understandably given the military origin of much early nuclear technology. Few people are aware of the number of lives saved by the medical uses of radionuclides produced in nuclear reactors, or associate radiation with other beneficial applications. There is little appreciation of the role that nuclear power is playing and could play to an increasing extent in helping to solve one of the world's most pressing future problems: the need for a rapid transition to non-fossil energy sources.

A further problem of perception results from the extremely high safety standards that are applied to the nuclear industry. Prevention of further accidents such as Chernobyl is clearly of the utmost importance. However, although the routine operations of nuclear plants contribute only a very small fraction to public radiation exposures, very large investments continue to be made to decrease already low discharges of radioactivity still further, implying that they constitute a significant risk. Radioactive waste disposal has to satisfy far more stringent safety criteria than most toxic waste disposals, even when the potential hazards involved are similar. No transport container for hazardous chemicals has ever been tested by driving a diesel train into it at one hundred miles an hour, as was done in the United Kingdom with a container for used nuclear fuel (which, unlike the train, remained intact), nor has a rocket propelled military fighter aircraft been crashed into part of a chemical plant, as was done in the United States to demonstrate the strength of a nuclear containment structure (which, again, survived essentially undamaged). The natural reaction to such high safety standards and exceptional tests is the perception that the risks must indeed be extreme to require such extreme measures.

Such factors result in some very wide divergences between perceived and technically assessed or measured risks. A study in the United States, for example, found that members of the League of Women Voters and college students ranked nuclear power as the most hazardous among 30 products and activities, including smoking, alcohol, motor vehicles and handguns, which between them result in over 300 000 deaths per year in the USA. In the same study, pesticides were ranked ninth (fourth by college students) although there is no evidence that their use resulted in any deaths at all, and medical X-rays were ranked 22nd (17th by college students) against a position of ninth on the basis of an estimated 2 300 deaths a year (on the linear no-threshold hypothesis).

Few people involved in the use or regulation of radiation would deny the importance of taking perceptions as well as technical assessments of risk into account in the decision-making process. What is less clear is the weight that should be attached to such perceptions. The cost of safety is usually paid by the public, either as consumer or as taxpayer. Excessive expenditure on making an already very safe activity even safer drains resources that could be better spent, either on improving less safe activities or on generally more beneficial measures such as better health care.

Public communication and consultation

For many years, the nuclear industry spent large sums on public information programmes aimed at improving the acceptability of nuclear power. The approach was often along the lines of "these are the facts about nuclear power and its benefits; if only more people were aware of them and understood them properly, nuclear power would become more acceptable."

More recently, there has been a growing realisation of the need to communicate with and consult the public, rather than simply to inform them. The public rightly demands that their perceptions and wishes are treated with respect and properly taken into account in the decision-making process. Risk communication is not a one-directional process, but one that should involve all the social actors: those who impose risks and are exposed to them, legislators, independent experts and regulators.

A key role in communication is played by the mass media. TV is particularly important, because it is the main source of information for many people. It is difficult for television to present a balanced perspective on a risk topic. Disasters are newsworthy, while safety is dull. Lively debates and arguments are interesting, while careful presentations of consensus opinions are less likely to retain viewers' attention. Discussions are usually confrontational and seldom reflect the scientific weight of the opposing views. Scientists tend to be careful and precise, exposing the uncertainties and limitations of knowledge as well as the consensus view. The representatives of pressure groups are highly skilled in attracting media attention, and their objectives and motivations are focused more on achieving their aims than on arriving at the truth or even at the best environmental option. Industry spokesmen, while generally more constrained by accountability requirements than pressure group representatives, also have very specific objectives.

A good example of the influence of the media is the question of clusters of childhood leukaemia near some nuclear installations. Intense media interest in such a cluster near the Sellafield reprocessing plant in the United Kingdom, and a series of similar claims elsewhere has led to a widespread public perception that nuclear installations in general are an important cause of childhood leukaemias. This possibility has been very carefully examined by a large number of independent experts and there is now an overwhelming scientific consensus that the observed increases were caused neither by discharges of radioactivity from the plants nor by radiation exposures of the fathers of the affected children, another widely reported suggestion. The most likely explanation is that the excesses are caused, at least in part, by cross-infections arising from population influx into an isolated rural area, an explanation supported by findings of similar clusters in other areas where such population mixing has occurred, remote from any nuclear installation. The media have largely ignored the careful scientific refutation of the original claims of causation of childhood leukaemia clusters by radiation, and the perception persists and is one of the reasons for strong opposition to any new nuclear siting proposals.

The Nuclear Energy Agency and other international and national bodies, as well as organisations representing the nuclear industry, are devoting increasing attention to achieving better communication and public participation in nuclear decision-making. The Swedish Risk Academy, in a recent report to the IAEA, concluded:

> "It is more and more recognised that coping with risks created by man's scientific and technological ingenuity is not only a technological but also a cultural, ethical and philosophical issue. These risks should be seen in the perspective of more general problems of development and human values. It is obvious that a world without risk cannot be achieved and that science and technology will continue to generate threats and not only benefits to humans. But there is still much place for making our decisions more thoughtful, enlightened and wise, and the relations between man, technology and the environment more bearable. In this context, risk communication is required to revise its risk-acceptability paradigm of promoting "understanding of technology" and instead aim at an understanding of the necessary compatibility between technological, environmental and social issues."

The cost of protection

In addition to the essentially social issues of ethics, risk perception and public involvement, the principles of radiation protection inevitably involve questions of the valuation of radiation detriment and the costs of the various methods used to protection of workers and the public, described in the previous chapter.

The "value" of a life

It is repugnant to assign a money value to human life or serious illness; no amount of money can compensate for the loss of a husband, wife or child, but collectively society often has to take decisions that involve such calculations, for example in allocating resources to health care or road safety improvements.

The amounts actually spent to "save" (strictly speaking, of course, to extend) one statistical life varies enormously, from very low figures when seeking to provide food or medical assistance to people in the world's poorest countries to many millions or sometimes even billions of dollars for some industrial regulatory actions. The ranges of values that can be inferred from investment and regulatory decisions for several sectors are shown in Table 4. The lowest inferred values are in the medical field and the highest in the chemical and nuclear industries.

Table 4: Inferred values of statistical life

SECTOR	RANGE OF INFERRED VALUES US$ 1986 (rounded) 95 percentile range
Medical	3 000 - 400 000
Road transport	3 000 - 2 800 000
Air transport	110 000 - 500 000
Other industry	400 000 - 15 000 000
Chemical industry	15 000 - 180 000 000
Nuclear industry	150 000 - 190 000 000

Source: Environmental Risk Assessment Unit, University of East Anglia, 1988

Another approach to deriving a "value of a life" is by determining what people are willing to pay for a reduction in risk, by asking them or observing the choices they make in practice, for example a willingness to accept a higher risk in return for higher pay. Again, the figures derived cover a wide range, from a few hundred thousand dollars to over ten million dollars, with a median value of about $2.5 million (1990 values).

Investment decisions for radiation protection purposes by the nuclear industry imply "values of life" that are typically a factor of ten higher than the median value of $2.5 million, or even more in some instances. For example, in the United Kingdom, investment in effluent clean-up equipment at Sellafield corresponds to an expenditure of around $400 million to prevent the statistical possibility, on the basis of the linear no-threshold hypothesis, of one or two cancer deaths in total over the next ten thousand years. In contrast, decisions relating to radiation protection in diagnostic radiology imply values of life between $10 000 and $100 000, and decisions of householders who have considered taking action on high radon levels in their homes imply values of life between $30 000 and $130 000.

Costing radiation exposure

The "cost" of a unit of radiation dose can, in principle, be simply derived from the "value of life" and ICRP's estimate of the risk of radiation. Using the median value of life of $2.5 million derived from studies of people's willingness to pay for risk reductions and the ICRP estimate for the risk of fatal cancer (at low doses and dose rates) gives a cost of a person-Sv of about $50 000.

Calculations such as this do not, of course, justify any particular figure being used for the value of a life-year or the cost of a person-Sv; they simply reflect the wide range of investment decisions that have been taken in different countries in many different circumstances. They do, however, suggest a level of expenditure against which future decisions can be judged. Cost-benefit analysis should not dominate decisions of optimisation of radiation protection, but it can indicate whether or not particular protection measures are likely to be worth implementing.

Costing future detriment

For some practices, radiation exposures may be delayed, sometimes for very long periods; this is particularly true in the case of radioactive waste disposals. It is common in cost-benefit analysis to discount future costs, on the basis that $1 now is worth more than $1 in the future, and discounting has sometimes been applied to future radiation detriment. Such discounting, however, implies that a life year saved now is worth more than a life year saved in the future. Whereas an individual may well value years of his or her own life saved now more highly than years saved in the future, it does not seem acceptable in general to appear to place a lower value on the lives of future generations than on those of our own generation. Indeed, one of the widely agreed principles of radioactive waste management is that the protection of future generations should be at least as good as the protection of the generation producing and disposing of the wastes.

Chapter 7

FUTURE PROBLEMS AND OPPORTUNITIES

Radiation protection recommendations have developed considerably over the years, reflecting society's demands for ever-increasing safety standards in general, progress in our understanding of radiation and its effects, and the growth in the number of applications of radiation in industry, health and scientific research. The recommendations will need to continue to develop as new social requirements, scientific findings and technologies emerge.

Problems may also arise when new recommendations are implemented. New protection measures need to be incorporated into the design and operation of facilities, processes which take time. There are also questions concerning the scope of the recommendations: does an employer need to take into account variations in background levels of radiation which will affect the overall dose to which his employees are exposed? As the issues surrounding radiation protection become more complex, it becomes increasingly important to develop recommendations that are simple, practicable, and can be readily understood.

Further consideration is also needed in relation to accident and chronic exposure situations. Although it is clear what action should be taken to protect the public directly after an accident, the current recommendations do not address the issue of timing the return from an accident situation to a normal system of radiation protection. Moreover, should interventions aim eventually to reduce chronic exposures to natural or man-made sources to the levels permitted to new practices? Work in these areas will need to take account of social and economic, as well as scientific factors.

There are many active scientific research programmes on radiation and its health effects. There is continuing debate on a number of fundamental issues such as the precise relationship between dose and risk, whether certain individuals are genetically susceptible to radiation, and whether it is possible to ascribe individual cases of cancer unambiguously to radiation exposure. Any of these areas of research may influence the development of future radiation protection practices.

Enormous progress has been made in understanding radiation and its health effects in the century since the pioneering discoveries of Röntgen and Becquerel. There has also been a dramatic growth in the number of beneficial applications, originally in the field of medicine, and increasingly, with the availability of a wide range of man-made radionuclides from nuclear reactors, throughout industry, agriculture and research. However, as in any other active scientific and technical field, areas of uncertainty remain, or appear as a result of new developments.

Developments in the regulatory area, too, can raise new problems. Full implementation of the 1977 recommendations of the ICRP, which introduced explicitly the three basic principles of justification, optimisation and limitation of exposure, took at least a decade. For example, it took many years of dialogue between those concerned with radiation protection and engineers and practitioners for optimisation of protection to be incorporated into facility design and operation, although in many cases optimisation is an implicit process based on the common sense requirement of making doses as low as reasonably achievable. The implementation of ICRP's 1990 recommendations, which introduced some new ideas, concepts and values, is raising a number of new problems.

The relationship between dose and risk

There is a continuing debate about the validity of the basic assumption on which modern radiation protection is based: a linear, non-threshold relationship between exposure and risk at the levels of dose currently received occupationally and in everyday life. Indeed there are many arguments in favour of other relationships, depending on the type of radiation, the type of risk, the distribution of exposure, etc. The non-threshold hypothesis has not been proven, nor have thresholds been observed, except in the case of deterministic effects at high doses.

A clear distinction is needed between radiobiology, a science, and radiation protection, which is an application of radiobiological knowledge to the protection of workers and the public and the management of risk. Elucidating the biological mechanisms at low doses is of considerable scientific interest and importance; in particular it might bring great progress in the understanding of carcinogenesis. However, the precise shape of the relationship between dose and risk is in general of limited relevance to radiation protection, where the primary need is for simple and practicable tools. For example, radiation risk management would be neither practicable nor feasible if the risk corresponding to a given dose varied with the amount of dose previously received, as would be the case on the basis of anything other than a linear relationship. The important issue for protection purposes is the safety margin that results from the use of the linear relationship, since most of the other possible relationships which are consistent with the epidemiological and experimental evidence result in lower risk estimates than those calculated from the linear relationship.

The threshold issue raises fundamental questions such as the importance of adaptive response to radiation exposure and the possibility of beneficial effects at low doses. Again these questions have little impact on the practices of protection; they would not influence the choice of dose limits, since any possible threshold would lie in the same region as the dose limits. The existence of beneficial effects, if proven, would not affect the fundamental principles of radiation protection, since harmful and beneficial effects, involving different mechanisms, may well coexist.

However, while arguments about linearity and thresholds should not affect the practice of radiation protection, they are of importance in assessing the consequences of exposures. The claims often made by anti-nuclear campaigners, that large numbers of cancers will be caused by exposures of large numbers of people to doses that are small fractions of annual natural background doses, are based on an unproven hypothesis.

The most important problems that arise from the latest ICRP recommendations relate to their practical implementation in particular circumstances, such as the management of occupational exposure to radon and chronic exposures of the public to man-made sources, and to the increasing difficulty of implementing increasingly complex and refined recommendations. These problems will concern the workforce and the public, routine operations, and the management of accidents and past situations.

Occupational exposures

Occupational exposure is currently defined as any exposure received at work, regardless of the source. This includes natural as well as man-made sources and therefore the problem of radon in the workplace must be considered. In practice, it will be up to national authorities to exclude from regulation those exposures that are not amenable to control. It will be necessary to define action levels, both for radon and for other natural sources, above which the exposure will be considered as occupational and subject to control under the general system of worker protection. Current recommended action levels for radon are in the range 3-10 mSv in a year. This implies that some installations, situated in regions with a high radon concentration, may be penalised when compared with similar installations in other areas.

Similar problems arise when workers are chronically exposed to residual contamination from past practices. This practical problem, encountered in old installations where past working habits did not comply with today's safety and protection standards, demonstrates the difficulty in applying some recommendations, which aim to provide the best available protection. This may result in apparent inequity, which would be difficult to explain and justify to the workforce.

Chronic exposure of the public

The public may be chronically exposed to radiation at levels high enough to justify considering remedial actions. The most important source of such exposure is radon. Despite this being by far the largest source of radiation exposure to the world's population, little has yet been done to reduce doses. In the OECD area alone, for example, there are about 2.5 million homes with radon concentration above the IAEA recommended action level involving about 7 million people. A very large collective dose could be averted through relatively simple remedial measures, at a cost per person-sievert much lower than the value currently used for radiation protection purposes. Therefore, provided that the assumption of a linear relationship between radon concentration in homes and the development of radiation-induced lung cancer does not significantly overestimate the risk, the number of cancers averted would justify the implementation of remedial measures. On the other hand, implementation would result in a waste of resources if the risk were significantly overestimated. This shows the difficulties in decision-making and the need for additional research on low dose effects. Similar problems arise when considering other sources of chronic exposure.

The problem of complexity

The examples illustrated above show that as the system of radiation protection has become more precise, it has also become more and more complex. This, together with the remaining uncertainties, explains why even those in the radiation protection community find difficulties and need time to become fully familiar with the system. This growing complexity, as well as the plethora of unfamiliar terms, units and definitions, might inhibit the proper use of the main principles of protection and contribute to the poor level of public understanding.

This would be particularly unfortunate in the case of a crisis, such as a nuclear accident, resulting in a lack of confidence by the public in the professionals working on their behalf.

Radiation remains a cause of public anxiety regardless of the effectiveness of measures which enable people to live with radiation in relative safety. All this has led to a greater awareness of the costs rather than the benefits of some radiation practices. This social dimension of radiation protection explains why there is a real need for some simplification. Furthermore, the system of radiation protection, which is substantially more refined than those for other hazardous materials, could play a central role in the development of a more integrated protection system. This would result in a better allocation of resources for protection, in general, by placing radiation risks in a more realistic perspective compared with other risks.

Scientific developments

There are substantial national and international research programmes on radiobiology and related topics, which are likely to contribute to improved radiation protection and might result in some breakthroughs in fundamental scientific knowledge. Some possible developments, although speculative and of an uncertain time-scale, can already be outlined and their implications explored.

Genetic susceptibility to radiation, already mentioned as an ethical issue, may affect employment policies and influence regulations, and a strategy for the future should be developed. There are two extreme approaches, both unacceptable on social grounds. The first would be to select workers on the basis of genetic predisposition; this might result in higher dose limits for workers since such selection could be used to identify individuals with above average radiation resistance. However, it would involve discriminating between workers and banning some workers from some jobs on the basis of genetic susceptibility to disease, which is not acceptable. The second approach would be to lower limits to a level at which the most sensitive individual would have the same protection as offered to the average individual under the current system of protection; this approach would effectively end almost all practices involving radiation, however beneficial, since it is thought that some rare individuals may be between ten and a hundred times more sensitive to radiation than normal.

A necessary first stage in developing an acceptable solution to this difficult problem is to establish the fraction of the population that is extremely sensitive to radiation. If this is only of the order of one in a thousand or one in ten thousand, the total risk to a given workforce, typically of hundreds of individuals, can be assessed with reasonable confidence. If the excess risk to the total workforce resulting from extreme genetic susceptibility is significant compared with the "normal" risk, some overall changes to work practices may be needed to bring the average risk down to acceptable levels. If the excess risk is insignificant, it may be considered acceptable to continue with the existing work practices, although some individuals may run higher risks than others exposed to the same levels of dose. This problem illustrates the sort of difficulties that may arise in the future, for which there is no clear solution today.

A related problem which may be more amenable to solution is the possible additive or synergistic effect of radiation and some diseases, especially those resulting in immune defence depression, such as virus infections. Since these diseases are generally curable or of short duration, this issue would only imply transient unsuitability for radiation work, without raising general and difficult ethical problems.

It may become possible to identify radiation-induced cancers among other "natural" cancers, through a specific "signature" that radiation would "print" at the molecular level. At present, the only scientific way to demonstrate the carcinogenic effect of radiation on humans is through epidemiological studies. These have their limitations: the proof is of a statistical nature, they can only offer a probability of causation, which is rarely higher than a few per cent, and they cannot, for statistical reasons, detect any significant effect below doses of about 50 mSv, which is well above natural background levels, typical occupational exposure levels, or even most doses to the public from extreme accidents such as Chernobyl. The development of a technique for identifying radiation induced cancers would effectively increase the power of epidemiology by a large margin. However, because of the large variety of cancer types in various organs and tissues, it is not expected that knowledge of this "mapping" will be complete for a long time, although new techniques may speed up the rate of progress. If such identification became possible for a wide enough range of cancers, especially those known to be easily induced by radiation, compensation for occupational disease could be based on a more solid basis than the probability of causation approach mainly used at present.

The need for further research

Most of the problems outlined above, which may have some impact on future radiation protection practices, will be solved only by intensive research, mainly involving radiobiology including, more especially, molecular biology.

The fundamental problem of quantifying the effects of low doses cannot be solved by observations of exposed populations and will need extended research, including fields other than radiation protection and even radiobiology.

This kind of research will impact mainly on the management of occupational exposures and little on public exposures, which are dominated by natural components which are not amenable to control (except radon to a certain degree), with generally trivial contributions from man-made sources.

Further research is needed in relation to accident situations, which might result in high worker doses with acute severe consequences and lower doses to the public, with the possibility of late effects (cancers and genetic defects). Research areas include the medical management of radiation casualties, particularly following large whole body exposures and combinations of radiation exposure with other injuries, and general health management to reduce public exposures.

The protection of the public in the case of major nuclear accidents is now well codified. Nevertheless, one remaining problem is the definition of the criteria for "return to normal life" following interventions, particularly highly disruptive ones such as evacuation. The dose limits that apply to practices are irrelevant in such situations; the criteria must be based on the predetermined intervention levels for accident situations. The accident situation can be considered to continue for as long as some form of intervention results in significant reduction of public exposure. After some time, depending on the natural processes of decay and dispersion of radioactivity and the effectiveness of clean-up operations, the benefits of the intervention in terms of reduced exposure become negligible or do not counterbalance the associated risks and costs. The fundamental problem which has been experienced in past severe accidents such as Chernobyl is the timing of the return from the accident situation, governed by intervention levels, to the normal system of radiation protection, governed by dose limits. Although this problem does not call for further scientific research, it requires deep reflection covering a wide range of disciplines including protection, economics, politics, psychology and demography.

A similarly complex issue which is not yet covered by the internationally agreed principles for radiation protection is the contribution of social factors to the effects of an accident. Operational aspects of radiation protection require an awareness of these socio-psychological factors, which apply equally to accidents not involving radiation exposure. Chernobyl has shown that the most severe effects of an accident may be due to stress, anxiety and a feeling of loss of control over one's circumstances, resulting from social and economic disruption and lack of confidence in the authorities, adding to the radiological consequences themselves. Once again, progress will be made only by combining the experience of specialists in many fields, especially those who have experience in conditions such as post traumatic stress disorder.

MAIN SALES OUTLETS OF OECD PUBLICATIONS
PRINCIPAUX POINTS DE VENTE DES PUBLICATIONS DE L'OCDE

AUSTRALIA – AUSTRALIE
D.A. Information Services
648 Whitehorse Road, P.O.B 163
Mitcham, Victoria 3132　Tel. (03) 9210.7777
　　　　　　　　　Fax: (03) 9210.7788

AUSTRIA – AUTRICHE
Gerold & Co.
Graben 31
Wien I　　　　　　Tel. (0222) 533.50.14
　　　　　　　Fax: (0222) 512.47.31.29

BELGIUM – BELGIQUE
Jean De Lannoy
Avenue du Roi, Koningslaan 202
B-1060 Bruxelles Tel. (02) 538.51.69/538.08.41
　　　　　　　　　Fax: (02) 538.08.41

CANADA
Renouf Publishing Company Ltd.
5369 Canotek Road
Unit 1
Ottawa, Ont. K1J 9J3　　Tel. (613) 745.2665
　　　　　　　　Fax: (613) 745.7660

Stores:
71 1/2 Sparks Street
Ottawa, Ont. K1P 5R1　Tel. (613) 238.8985
　　　　　　　　Fax: (613) 238.6041

12 Adelaide Street West
Toronto, QN M5H 1L6　Tel. (416) 363.3171
　　　　　　　　Fax: (416) 363.5963

Les Éditions La Liberté Inc.
3020 Chemin Sainte-Foy
Sainte-Foy, PQ G1X 3V6 Tel. (418) 658.3763
　　　　　　　　Fax: (418) 658.3763

Federal Publications Inc.
165 University Avenue, Suite 701
Toronto, ON M5H 3B8　Tel. (416) 860.1611
　　　　　　　　Fax: (416) 860.1608

Les Publications Fédérales
1185 Université
Montréal, QC H3B 3A7　Tel. (514) 954.1633
　　　　　　　　Fax: (514) 954.1635

CHINA – CHINE
Book Dept., China Natinal Publications
Import and Export Corporation (CNPIEC)
16 Gongti E. Road, Chaoyang District
Beijing 100020　Tel. (10) 6506-6688 Ext. 8402
　　　　　　　　　(10) 6506-3101

CHINESE TAIPEI – TAIPEI CHINOIS
Good Faith Worldwide Int'l. Co. Ltd.
9th Floor, No. 118, Sec. 2
Chung Hsiao E. Road
Taipei　　　　Tel. (02) 391.7396/391.7397
　　　　　　　　Fax: (02) 394.9176

**CZECH REPUBLIC –
RÉPUBLIQUE TCHÈQUE**
National Information Centre
NIS – prodejna
Konviktská 5
Praha 1 – 113 57　　Tel. (02) 24.23.09.07
　　　　　　　　Fax: (02) 24.22.94.33
E-mail: nkposp@dec.niz.cz
Internet: http://www.nis.cz

DENMARK – DANEMARK
Munksgaard Book and Subscription Service
35, Nørre Søgade, P.O. Box 2148
DK-1016 København K　Tel. (33) 12.85.70
　　　　　　　　Fax: (33) 12.93.87

J. H. Schultz Information A/S,
Herstedvang 12,
DK – 2620 Albertslung　Tel. 43 63 23 00
　　　　　　　　Fax: 43 63 19 69
Internet: s-info@inet.uni-c.dk

EGYPT – ÉGYPTE
The Middle East Observer
41 Sherif Street
Cairo　　　　　　Tel. (2) 392.6919
　　　　　　　　Fax: (2) 360.6804

FINLAND – FINLANDE
Akateeminen Kirjakauppa
Keskuskatu 1, P.O. Box 128
00100 Helsinki

Subscription Services/Agence d'abonnements :
P.O. Box 23
00100 Helsinki　　　Tel. (358) 9.121.4403
　　　　　　　　Fax: (358) 9.121.4450

*FRANCE
OECD/OCDE
Mail Orders/Commandes par correspondance :
2, rue André-Pascal
75775 Paris Cedex 16 Tel. 33 (0)1.45.24.82.00
　　　　　　Fax: 33 (0)1.49.10.42.76
　　　　　　　　Telex: 640048 OCDE
Internet: Compte.PUBSINQ@oecd.org

Orders via Minitel, France only/
Commandes par Minitel, France
exclusivement : 36 15 OCDE

OECD Bookshop/Librairie de l'OCDE :
33, rue Octave-Feuillet
75016 Paris　　　　Tel. 33 (0)1.45.24.81.81
　　　　　　　　33 (0)1.45.24.81.67

Dawson
B.P. 40
91121 Palaiseau Cedex　Tel. 01.89.10.47.00
　　　　　　　　Fax: 01.64.54.83.26

Documentation Française
29, quai Voltaire
75007 Paris　　　　Tel. 01.40.15.70.00

Economica
49, rue Héricart
75015 Paris　　　　Tel. 01.45.78.12.92
　　　　　　　　Fax: 01.45.75.05.67

Gibert Jeune (Droit-Économie)
6, place Saint-Michel
75006 Paris　　　　Tel. 01.43.25.91.19

Librairie du Commerce International
10, avenue d'Iéna
75016 Paris　　　　Tel. 01.40.73.34.60

Librairie Dunod
Université Paris-Dauphine
Place du Maréchal-de-Lattre-de-Tassigny
75016 Paris　　　　Tel. 01.44.05.40.13

Librairie Lavoisier
11, rue Lavoisier
75008 Paris　　　　Tel. 01.42.65.39.95

Librairie des Sciences Politiques
30, rue Saint-Guillaume
75007 Paris　　　　Tel. 01.45.48.36.02

P.U.F.
49, boulevard Saint-Michel
75005 Paris　　　　Tel. 01.43.25.83.40

Librairie de l'Université
12a, rue Nazareth
13100 Aix-en-Provence　Tel. 04.42.26.18.08

Documentation Française
165, rue Garibaldi
69003 Lyon　　　　Tel. 04.78.63.32.23

Librairie Decitre
29, place Bellecour
69002 Lyon　　　　Tel. 04.72.40.54.54

Librairie Sauramps
Le Triangle
34967 Montpellier Cedex 2 Tel. 04.67.58.85.15
　　　　　　　　Fax: 04.67.58.27.36

A la Sorbonne Actual
23, rue de l'Hôtel-des-Postes
06000 Nice　　　　Tel. 04.93.13.77.75
　　　　　　　　Fax: 04.93.80.75.69

GERMANY – ALLEMAGNE
OECD Bonn Centre
August-Bebel-Allee 6
D-53175 Bonn　　　Tel. (0228) 959.120
　　　　　　　　Fax: (0228) 959.12.17

GREECE – GRÈCE
Librairie Kauffmann
Stadiou 28
10564 Athens　　　Tel. (01) 32.55.321
　　　　　　　　Fax: (01) 32.30.320

HONG-KONG
Swindon Book Co. Ltd.
Astoria Bldg. 3F
34 Ashley Road, Tsimshatsui
Kowloon, Hong Kong　　Tel. 2376.2062
　　　　　　　　Fax: 2376.0685

HUNGARY – HONGRIE
Euro Info Service
Margitsziget, Európa Ház
1138 Budapest　　　Tel. (1) 111.60.61
　　　　　　　　Fax: (1) 302.50.35
E-mail: euroinfo@mail.matav.hu
Internet: http://www.euroinfo.hu//index.html

ICELAND – ISLANDE
Mál og Menning
Laugavegi 18, Pósthólf 392
121 Reykjavik　　　Tel. (1) 552.4240
　　　　　　　　Fax: (1) 562.3523

INDIA – INDE
Oxford Book and Stationery Co.
Scindia House
New Delhi 110001　Tel. (11) 331.5896/5308
　　　　　　　　Fax: (11) 332.2639
E-mail: oxford.publ@axcess.net.in

17 Park Street
Calcutta 700016　　　　Tel. 240832

INDONESIA – INDONÉSIE
Pdii-Lipi
P.O. Box 4298
Jakarta 12042　　　Tel. (21) 573.34.67
　　　　　　　　Fax: (21) 573.34.67

IRELAND – IRLANDE
Government Supplies Agency
Publications Section
4/5 Harcourt Road
Dublin 2　　　　　Tel. 661.31.11
　　　　　　　　Fax: 475.27.60

ISRAEL – ISRAËL
Praedicta
5 Shatner Street
P.O. Box 34030
Jerusalem 91430　　Tel. (2) 652.84.90/1/2
　　　　　　　　Fax: (2) 652.84.93

R.O.Y. International
P.O. Box 13056
Tel Aviv 61130　　　Tel. (3) 546 1423
　　　　　　　　Fax: (3) 546 1442
E-mail: royil@netvision.net.il

Palestinian Authority/Middle East:
INDEX Information Services
P.O.B. 19502
Jerusalem　　　　Tel. (2) 627.16.34
　　　　　　　　Fax: (2) 627.12.19

ITALY – ITALIE
Libreria Commissionaria Sansoni
Via Duca di Calabria, 1/1
50125 Firenze　　　Tel. (055) 64.54.15
　　　　　　　　Fax: (055) 64.12.57
E-mail: licosa@ftbcc.it

Via Bartolini 29
20155 Milano　　　Tel. (02) 36.50.83

Editrice e Libreria Herder
Piazza Montecitorio 120
00186 Roma　　　　Tel. 679.46.28
　　　　　　　　Fax: 678.47.51

Libreria Hoepli
Via Hoepli 5
20121 Milano   Tel. (02) 86.54.46
   Fax: (02) 805.28.86

Libreria Scientifica
Dott. Lucio de Biasio 'Aeiou'
Via Coronelli, 6
20146 Milano   Tel. (02) 48.95.45.52
   Fax: (02) 48.95.45.48

JAPAN – JAPON
OECD Tokyo Centre
Landic Akasaka Building
2-3-4 Akasaka, Minato-ku
Tokyo 107   Tel. (81.3) 3586.2016
   Fax: (81.3) 3584.7929

KOREA – CORÉE
Kyobo Book Centre Co. Ltd.
P.O. Box 1658, Kwang Hwa Moon
Seoul   Tel. 730.78.91
   Fax: 735.00.30

MALAYSIA – MALAISIE
University of Malaya Bookshop
University of Malaya
P.O. Box 1127, Jalan Pantai Baru
59700 Kuala Lumpur
Malaysia   Tel. 756.5000/756.5425
   Fax: 756.3246

MEXICO – MEXIQUE
OECD Mexico Centre
Edificio INFOTEC
Av. San Fernando no. 37
Col. Toriello Guerra
Tlalpan C.P. 14050
Mexico D.F.   Tel. (525) 528.10.38
   Fax: (525) 606.13.07
E-mail: ocde@rtn.net.mx

NETHERLANDS – PAYS-BAS
SDU Uitgeverij Plantijnstraat
Externe Fondsen
Postbus 20014
2500 EA's-Gravenhage   Tel. (070) 37.89.880
Voor bestellingen: Fax: (070) 34.75.778

Subscription Agency/Agence d'abonnements :
SWETS & ZEITLINGER BV
Heereweg 347B
P.O. Box 830
2160 SZ Lisse   Tel. 252.435.111
   Fax: 252.415.888

NEW ZEALAND –
NOUVELLE-ZÉLANDE
GPLegislation Services
P.O. Box 12418
Thorndon, Wellington   Tel. (04) 496.5655
   Fax: (04) 496.5698

NORWAY – NORVÈGE
NIC INFO A/S
Ostensjoveien 18
P.O. Box 6512 Etterstad
0606 Oslo   Tel. (22) 97.45.00
   Fax: (22) 97.45.45

PAKISTAN
Mirza Book Agency
65 Shahrah Quaid-E-Azam
Lahore 54000   Tel. (42) 735.36.01
   Fax: (42) 576.37.14

PHILIPPINE – PHILIPPINES
International Booksource Center Inc.
Rm 179/920 Cityland 10 Condo Tower 2
HV dela Costa Ext cor Valero St.
Makati Metro Manila   Tel. (632) 817 9676
   Fax: (632) 817 1741

POLAND – POLOGNE
Ars Polona
00-950 Warszawa
Krakowskie Prezdmiescie 7   Tel. (22) 264760
   Fax: (22) 265334

PORTUGAL
Livraria Portugal
Rua do Carmo 70-74
Apart. 2681
1200 Lisboa   Tel. (01) 347.49.82/5
   Fax: (01) 347.02.64

SINGAPORE – SINGAPOUR
Ashgate Publishing
Asia Pacific Pte. Ltd
Golden Wheel Building, 04-03
41, Kallang Pudding Road
Singapore 349316   Tel. 741.5166
   Fax: 742.9356

SPAIN – ESPAGNE
Mundi-Prensa Libros S.A.
Castelló 37, Apartado 1223
Madrid 28001   Tel. (91) 431.33.99
   Fax: (91) 575.39.98
E-mail: mundiprensa@tsai.es
Internet: http://www.mundiprensa.es

Mundi-Prensa Barcelona
Consell de Cent No. 391
08009 – Barcelona   Tel. (93) 488.34.92
   Fax: (93) 487.76.59

Libreria de la Generalitat
Palau Moja
Rambla dels Estudis, 118
08002 – Barcelona
   (Suscripciones) Tel. (93) 318.80.12
   (Publicaciones) Tel. (93) 302.67.23
   Fax: (93) 412.18.54

SRI LANKA
Centre for Policy Research
c/o Colombo Agencies Ltd.
No. 300-304, Galle Road
Colombo 3   Tel. (1) 574240, 573551-2
   Fax: (1) 575394, 510711

SWEDEN – SUÈDE
CE Fritzes AB
S–106 47 Stockholm   Tel. (08) 690.90.90
   Fax: (08) 20.50.21

For electronic publications only/
Publications électroniques seulement
STATISTICS SWEDEN
Informationsservice
S-115 81 Stockholm   Tel. 8 783 5066
   Fax: 8 783 4045

Subscription Agency/Agence d'abonnements :
Wennergren-Williams Info AB
P.O. Box 1305
171 25 Solna   Tel. (08) 705.97.50
   Fax: (08) 27.00.71

Liber distribution
Internatial organizations
Fagerstagatan 21
S-163 52 Spanga

SWITZERLAND – SUISSE
Maditec S.A. (Books and Periodicals/Livres
et périodiques)
Chemin des Palettes 4
Case postale 266
1020 Renens VD 1   Tel. (021) 635.08.65
   Fax: (021) 635.07.80

Librairie Payot S.A.
4, place Pépinet
CP 3212
1002 Lausanne   Tel. (021) 320.25.11
   Fax: (021) 320.25.14

Librairie Unilivres
6, rue de Candolle
1205 Genève   Tel. (022) 320.26.23
   Fax: (022) 329.73.18

Subscription Agency/Agence d'abonnements :
Dynapresse Marketing S.A.
38, avenue Vibert
1227 Carouge   Tel. (022) 308.08.70
   Fax: (022) 308.07.99

See also – Voir aussi :
OECD Bonn Centre
August-Bebel-Allee 6
D-53175 Bonn (Germany)   Tel. (0228) 959.120
   Fax: (0228) 959.12.17

THAILAND – THAÏLANDE
Suksit Siam Co. Ltd.
113, 115 Fuang Nakhon Rd.
Opp. Wat Rajbopith
Bangkok 10200   Tel. (662) 225.9531/2
   Fax: (662) 222.5188

TRINIDAD & TOBAGO, CARIBBEAN
TRINITÉ-ET-TOBAGO, CARAÏBES
Systematics Studies Limited
9 Watts Street
Curepe
Trinidad & Tobago, W.I.   Tel. (1809) 645.3475
   Fax: (1809) 662.5654
E-mail: tobe@trinidad.net

TUNISIA – TUNISIE
Grande Librairie Spécialisée
Fendri Ali
Avenue Haffouz Imm El-Intilaka
Bloc B 1 Sfax 3000   Tel. (216-4) 296 855
   Fax: (216-4) 298.270

TURKEY – TURQUIE
Kültür Yayinlari Is-Türk Ltd.
Atatürk Bulvari No. 191/Kat 13
06684 Kavaklidere/Ankara
   Tel. (312) 428.11.40 Ext. 2458
   Fax : (312) 417.24.90

Dolmabahce Cad. No. 29
Besiktas/Istanbul   Tel. (212) 260 7188

UNITED KINGDOM – ROYAUME-UNI
The Stationery Office Ltd.
Postal orders only:
P.O. Box 276, London SW8 5DT
Gen. enquiries   Tel. (171) 873 0011
   Fax: (171) 873 8463

The Stationery Office Ltd.
Postal orders only:
49 High Holborn, London WC1V 6HB
Branches at: Belfast, Birmingham, Bristol,
Edinburgh, Manchester

UNITED STATES – ÉTATS-UNIS
OECD Washington Center
2001 L Street N.W., Suite 650
Washington, D.C. 20036-4922
   Tel. (202) 785.6323
   Fax: (202) 785.0350
Internet: washcont@oecd.org

Subscriptions to OECD periodicals may also
be placed through main subscription agencies.

Les abonnements aux publications périodiques
de l'OCDE peuvent être souscrits auprès des
principales agences d'abonnement.

Orders and inquiries from countries where Distributors have not yet been appointed should be
sent to: OECD Publications, 2, rue André-Pascal, 75775 Paris Cedex 16, France.

Les commandes provenant de pays où l'OCDE
n'a pas encore désigné de distributeur peuvent
être adressées aux Éditions de l'OCDE, 2, rue
André-Pascal, 75775 Paris Cedex 16, France.

12-1996

OECD PUBLICATIONS, 2, rue André-Pascal, 75775 PARIS CEDEX 16
PRINTED IN FRANCE
(66 97 04 1 P) ISBN 92-64-15483-3 – No. 49437 1997